IEE HISTORY OF TECHNOLOGY SERIES 11

Series Editor: Dr B. Bowers

METRES *to* MICROWAVES

British development of active components for radar systems 1937 to 1944

METRES *to* MICROWAVES

British development of active components for radar systems 1937 to 1944

E. B. Callick, BSc, FIEE

Peter Peregrinus Ltd. on behalf of the Institution of Electrical Engineers

Published by: Peter Peregrinus Ltd., London, United Kingdom

Peter Peregrinus Ltd.,
Michael Faraday House,
Six Hills Way, Stevenage,
Herts. SG1 2AY, United Kingdom

British Library Cataloguing in Publication Data

Callick, E. B. (Eric Brian)
    Metres to microwaves: British development of active
    components for radar systems.
    1. Radar. Development, history
    I. Title
    621.3848'09

ISBN 0 86341 212 2

Printed in England by Short Run Press Ltd., Exeter

Contents

Foreword
by
Professor Sir Alan Hodgkin, O.M., F.R.S.

As one of the early research workers in the field of airborne centimetric radar, it is a pleasure for me to write a foreword to Brian Callick's account of the special devices which made possible the transition from 'Metres to Microwaves'. There is little doubt that this revolution in technology, which took place between 1939 and 1945, was greatly accelerated by the urgent need to defend this country from bombing by night as well as by day.

To begin with, the pressure came primarily from the Royal Air Force, who wanted an Air Interception Radar that would enable them to detect Night Raiders or Minelayers at a range greater than the height above ground. To do this, a powerful source of radiation in the range 5–10 centimetres was needed, since there was no room inside a night fighter for a paraboloid bigger than about 70 centimetres. At first, the possibility of working at such short wave lengths seemed remote, but the development of the Randall-Boot magnetron in Birmingham and Megaw's work at the GEC Laboratories, which turned it into a practical proposition, gave centimetric radar a reality that it lacked previously. At the same time the potential applications to antisubmarine warfare and gunlaying began to be considered by the Admiralty and Army radar institutes.

However, before the magnetron could be used, some major problems in high frequency engineering had to be solved and it is with these, as well as with the development of the magnetron itself that Callick's book is concerned. It turned out that the best type of receiver was a crystal mixer using a reflexion klystron or Sutton tube as a local oscillator; this was followed by an intermediate frequency amplifier of conventional design. The names of H.W.B. Skinner and A.G. Ward are closely associated with this development, along with those of the inventor of the reflexion klystron, R.W. Sutton, and L.F. Broadway of E.M.I. who was responsible for its final development and large-scale production.

Finally, it was desirable in some applications, and essential in others, to be able to transmit and receive from one aerial. As Callick shows, the solution to this problem lay in the use of one, or more often two, gas filled Sutton tubes which flashed over when the magnetron was driven, but not when it was cold. This idea was proposed by A.H. Cooke of the Clarendon Laboratory and a practical device was built for the Air Interception application to the design of Ward, Skinner and Starr in the early summer of 1941.

Callick points out that these quite new problems, as well as many others of a more conventional kind, were solved remarkably quickly. Thus the Navy had a 9 centimetre Type 271 set at sea by mid-1941 and in December of the same year a Beaufighter equipped with a centimetric Air Interception set (Mark VII) had chased and damaged an enemy mine-layer.

These early operational successes stimulated the development of many military and navigational applications of 9 and 3 centimetre radar and resulted in a proliferation of the special devices required in this type of work.

Brian Callick's text and illustrations provide an enjoyable summary of these developments and show how they relate to a fascinating piece of military history.

A.L.H.

Preface

'Metres to Microwaves' is essentially a short history of British development of transmitting and receiving valves, display tubes, and other active components for radar systems used in the second world war.

Part 1 outlines the early history of CVD, the organisation responsible for research and development of active devices for use in Navy, Army and Air Force radar and communications equipment. The wording was chosen to make maximum use of contemporary records. Many of these are quoted verbatim. At the time of writing, no records have been found of the meetings of the CVD Policy and Technical Committees after July 1941. By that time, the structure of CVD was well established. The activities of the senior committees, including the choice of projects and the priorities allocated thereto, were reflected in the agenda of the specialist sub-committees, so that the history of CVD thereafter is essentially that of the active devices created by its members. This is the subject of Part 2. To assist readers without specialist knowledge, it includes an introduction to radar technology and outlines of two typical wartime systems. It also gives brief non-mathematical descriptions of the basic principles of operation of the devices which determined their performance. [Some of the critical active components were replaced many years ago by alternatives with greatly superior performance. For example, low noise signal amplifier valves were replaced by semiconductor devices, cathode ray tubes with afterglow screens were replaced by three-colour tubes driven by computer data storage systems.] Appendix 6 describes materials and technology used in the making of the devices described in chapters 3 to 6. Much of this had to be developed in parallel with design of the devices themselves.

Because it is a history of development, the terminology throughout the book is by choice that in common use at the time. Millibars were not then used as units of pressure; dimensions in inches were not always converted to millimetres (some drawings used both). British transmitters and receivers used 'electronic valves'; American equipment used 'electron tubes'.

As far as possible, the text is based on written records, preferably from more than one source. Where no other data has been located, first hand recollection has been confirmed by consultations with others associated with that particular activity. In a book of this size, it has not been possible to identify by name all the engineers and scientists who made individual contributions to the research and development which became the foundation of British radar technology, or to describe every product of their ingenuity, and so choices had to be made when describing the patterns of development of each family of active devices. Thus no account has been given of important indirect contributions to their development by designers of equipment, notably A.T. Starr of TRE, who worked with H.W.B. Skinner and A.G. Ward on the first airborne duplexer system, and W.T. Davies, B.W. Lythall and C. Domb of Signal School, who identified and quantified the effect of long transmission lines on the frequency stability of magnetrons. This led to modifications of their output circuits to achieve a compromise between available power output and the adverse effects of reflections in the tranmission line or waveguide between the magnetron and the aerial system, or from the radiating element itself, and to the introduction of 'pulling figure' into magnetron specifications.

For the same reason, no reference is made to theoretical studies such as those on the magnetron by D.R. Hartree and E.C. Stoner, on the principles of velocity modulation by J.H. Fremlin, A.W. Gent, D.P.R. Petrie, S.G. Tomlin and P.J. Wallace, and on the geometry of electrode assemblies used for the production of electron beams in cathode ray tubes and klystrons by O. Klemperer.

Detailed technical data has been abstracted from J.I.E.E. Vol 93 Part IIIA, 1946, and from wartime and early catalogues (EVS handbooks) which were the source of many of the outline drawings. To preserve the original format, most of these are reproduced as found. In doing so, it was accepted that the necesssary reduction in size would sometimes result in loss of clarity.

In gathering material for the text, I have been given valuable assistance by colleagues and friends, to whom I express my thanks and appreciation. It has been my good fortune to find a large part of the story in archives made available by Miss Janet Dudley, Senior Librarian at the Royal Signals and Radar Establishment, and her deputy, Mr. Peter Trevitt, who not only assembled a large amount of historic material but also provided easy access to it, for which I am particularly grateful. (The archives are held in the Public Record Office under the general classification Avia 7). Malcolm Foley was the source of almost all the information on development of silica transmitting and modulator valves at Signal school. His contribution was all the more valuable because of his detailed knowledge of the design and fabrication techniques used in their construction. He also provided much of the data for the chapter on general purpose valves. Robert Sutton wrote his own account of the early days of the reflex klystron. Through the courtesy of John Coales, whose recollections of the early days of radar and comments on events thereafter were very much more than a source of information, I was privileged to study the private scientific papers of the late Dr E.C.S. Megaw. I have also had the pleasure and benefit of conversations with Professor James Sayers, Mr L. Rushforth and Mr W.E. Willshaw, who were directly involved in research and development of the magnetron, with Professor Brebis Bleaney, of the Clarendon Laboratory, with Sir Charles Oatley, who was a member of the group which developed Army radar systems at Bawdsey, and with the late Dr. C.N. Smythe, leader of the team which developed grounded grid signal amplifiers for metric radar systems. Sir Robert Clayton provided copies of GEC reports on their wartime development of transmitting and receiving tubes for metric and microwave radar, and made valuable additions to data gathered from other sources. Through Mr Jim Lodge of Thorn-EMI, I had access to archival material which included data on the history of reflex klystrons. Information from Mr. Don Tomlin on the early days of anti-aircraft microwave radar complemented and illuminated that available from other records of magnetron development. Mr H.J.W. ('diode') Reeves provided information on the development of metric duplexer components at Signal School

My thanks are due to Paul Wright for his detailed scrutiny of the text and his constructive criticism of its content

Finally, I have to thank Robert Sutton for his assistance throughout the many months we worked together on preparation of the text. Apart from the labour of putting the whole of the manuscript through his word processor, his knowledge and understanding of the subject were invaluable.

E.B.C.

Abbreviations

Titles

D.C.D. — Director of Communications Development
D.D.C.D. — Deputy Director of Communications Development
D.S.R. — Director of Scientific Research, head of
(D).S.R.E. — Directorate of Scientific Research
    and Experiment, Admiralty
I.S.T.V.C. (T.V.C.) — Interservices Technical Valve
    Committee
I.S.V.P.C. — Interservices Valve Production Committee
M.A.P. — Ministry of Aircraft Production
M.O.S. — Ministry of Supply

Establishments

A.D.E.E.	Air Defence Experimental Establishment (part of War Department)
A.D.R.D.E.	Air Defence Research and Development Establishment, formerly A.D.E.E.
A.M.R.E.	Air Ministry Research Establishment, successor to 'Orfordness' and Bawdsey Research Station
A.S.E.	Admiralty Signals Establishment, formerly H.M. Signal School
A.S.R.E.	Admiralty Signals and Radar Establishment, formerly A.S.E.
B.R.S.	Bawdsey Research Station
R.R.D.E.	Radar Research and Development Establishment, formerly A.D.R.D.E. (part of Ministry of Supply), now part of R.S.R.E.
R.S.R.E.	Royal Signals and Radar Establishment
T.R.E.	Telecommunications Research Establishment, formerly A.M.R.E., now part of R.S.R.E.

Companies and University Laboratories

B.T.H.* — British Thomson Houston Ltd
B.T.L.[†] — Bell Telephone Laboratories,
      New York, U.S.A.
Birmingham University

C.G.T.S.F. — Compagnie Generale de Telegraphie sans Fil
      Paris
CinTel — Cinema Television Ltd
Clarendon Laboratory, Oxford
Columbia University, New York, U.S.A.
Cosmos Manufacturing Co Ltd*
Cossor, A.C. Ltd.

Ediswan* — Edison Swan Electrical Co. Ltd
E.E.V. — English Electric Valve Company Ltd
Ekco — E.K. Cole Ltd
E.M.I. — Electrical and Musical Industries Ltd

G.E.C. — General Electric Company Ltd

M.I.T. — Massachusetts Institute of Technology
      Cambridge, Mass, U.S.A.
M.O.V. — Marconi Osram Valve Co. Ltd.
Marconi's Wireless Telegraph Co. Ltd.
MetVick* — Metropolitan Vickers (Electrical Co.) Ltd.
Mullard — Mullard Radio Valve Company Ltd

Owen Harries, J.H.

Philips — N.V. Philips Gloeilampenfabriken,
      Eindhoven, Holland

RadLab — Radiation Laboratory of M.I.T.
R.C.A. — Radio Corporation of America
      Harrison, N.J., U.S.A.
Raytheon Corporation, Waltham, Mass., U.S.A.

S.F.R. — Societe Francaise Radioelectrique, Paris
S.T.C. — Standard Telephones and Cables Ltd.
Scophony Ltd.
Stanford, (The Leland), University
      Palo Alto, California, U.S.A.
Sylvania, (Hygrade), Emporium, Pa., U.S.A.

Western Electric,[†] Allentown, Pa., U.S.A.

    *[†] Associated Companies

| YEAR | TRANSMITTER VALVES | | MAGNETRONS | | RECEIVING VALVES | | REFLEX KLYSTRONS | | MIXERS |
	40–200	600 MHz	S-band	X-band	40–200	600 MHz	S-band	X-band	
1937	NT60								
1938	VT58								
1939	NT57T NT77 NT86 VT98 VT114	MICRO-PUPS ↓ VT90							
1940			Birmingham Xptl E11 89 NT 98 } CV 38 }		RL7		Signal School Xptl		
1941		NT99	Strapping CV41 CV56	CV 108	RL37	S25A (A) CV16 CV82 (LO)	NR89 CV35		TRE Xptl CV101 (X-Xptl) CV111
1942			CV120 CV76 CV64					CV87	CV253 CV58 (Diode)
1943		CV288	CV192	CV209 CV208		CV90 (LO)	CV116	CV129 CV130 CV322 CSS2	
1944			CV160	CV214 Clarendon 'Rising sun'					

DUPLEXER TUBES			DISPLAY TUBES		MODULATOR SWITCHES			YEAR
METRIC	S-band	X-band	'A-scope'	PPI	HARD VALVE	THYRATRON	TRIGATRON	
								1937
			VCR 84					1938
				'Afterglow'	NT78A			1939
			VCR 85					
				VCR 140A	NT100	CV13		1940
			CV1097	CV1140	CV44			
'TV' Diode								
CV86					CV57			
	Clarendon Xptl			CV1516		CV1145		1941
			CV1587					
						CV22		
E1421 CV8		TRE Xptl					CV85	
			CV964					1942
			CV1522	CV1517				
	CV43	CV114				CV12		
	CV210						CV125	
		CV221		CV1520				1943
		CV115						
	CV294							
				CV1530	CV313			1944
CV94								

Part 1

'C V D'

1.1 The Origins of CVD

In July 1935, an Air Ministry scientific group developed and tested satisfactorily at Orfordness a simple pulsed radio-location system operating at six metres wavelength (50 MHz). This wavelength was chosen partly because transmissions could be mistaken for a continuation of earlier work on reflections from the ionosphere, and partly because components for the transmitter and receiver were available from companies in the communications industry. A cathode ray oscilloscope used for electrical measurements was adapted to display the signal returning from target ships and aircraft. For reasons of high security the working group was small (six in all) and suppliers of equipment and components were kept completely unaware of the nature of the experiments.

In May 1936, the group moved to Bawdsey Manor, where they were joined by a similar group from the War Department, who concentrated on the development of mobile equipment and gun-laying systems. By the end of March 1937, the teams had designed, built and handed over to the RAF an early warning system which was the first of twenty in the coastal defence network Chain Home (CH). This operated at 13 metres (26 MHz) to minimise the probability of spurious returns from the atmosphere and interference from broadcast transmissions and other man-made radiation. The transmitting aerial was a vertical array of horizontal dipoles suspended between two masts three hundred feet high. Objects within the flooded target area were located by a conventional direction finding (DF) receiver with its own directional aerial. (This may be the reason why British radar in its early days was referred to as 'RDF'. The term 'radar' was coined by the US Services in 1943, and was ultimately used worldwide). The transmitting valves were supplied by the Admiralty Signal School, who had for many years developed and manufactured silica envelope valves for naval communications, and had been in touch with work at Bawdsey from the beginning. In July 1937, secret contracts were placed with Metropolitan Vickers Ltd. for the transmitters and A.C.Cossor Ltd. for the receivers and display units. The same contractors also supplied gun-laying equipment, GL1, to the Army. As before, other commercial suppliers of equipment and components were kept unaware of their intended use. Signal School continued to develop and supply transmitting valves, some of which were replaced later by demountable units designed and made by MetVick, and by metal-glass envelope valves developed under 'CVD' contract (see p. 3) at the Research Laboratories of the General Electric Company.

Development of the Chain Home system at the Bawdsey Research Station was followed by research and development of equipment operating at 200 MHz (1.5 m) for use in Chain Home Low (CHL) which provided improved low-level coverage for the coastal defence network. The teams also developed 200MHz mobile and aircraft RDF systems – GL2 (the MkII gun-laying equipment for the Army), ASV (Air to Surface Vessel), and AI (Air Interception), where the equipment had to be as small as practicable, and target bearings measured with an accuracy greater than that available from GL1. The early AI and ASV transmitters used Western Electric 316A 'doorknob' valves. Later versions, as well as GL2 and CHL, which were all operational by early 1940, used 'micropup' valves, developed under the 'CVD' contract by GEC from a type used later in TV transmitters. The receivers used American RCA 'acorns' for the signal amplifier and mixer.

Although Admiralty and Signal School were aware of the work at Bawdsey from the beginning, it was considered that the design and construction of ship-borne installations was sufficiently different from land-based or aircraft equipment to justify an independent approach to their development. By October 1938, two Type 79 ship-borne early warning systems operating at 43 MHz (7 m) had been installed and tested at sea. The frequency was chosen as a compromise between obtaining the maximum peak power (70 kW) from valves made by Signal School, and reducing the dimensions of the aerial so that it could withstand wind forces and be rotated by remote control from the 'office'. Its successor, Type 281, operated at 90 MHz (3.3 m) and was designed to provide longer range and increased accuracy of bearing measurement with an aerial system no larger than that of Type 79. The transmitter used valves developed and manufactured by Signal School to give peak powers up to one megawatt. Like Chain Home , Type 79 and Type 281 were designed to provide protection against attacks by hostile aircraft. Early in 1938, J.F.Coales and his group at Signal School had decided that the best prospect for detecting surface vessels would be to operate at 600 MHz (50 cm), it being estimated that this was the highest frequency at which triode valves could be pulsed to give high peak power output. This resulted in the development of several 600 MHz Naval equipments, including the Type 284 fire control system which was operational in late 1940. The transmitters used 'micropup' valves developed by GEC. Admiralty had at this time a contract with GEC for the research and development of communications equipment in this band, including a study of propagation by E.C.S. Megaw, and research by his group on magnetrons which could be used as transmitters. This Ultra Short Wave Admiralty contract also covered work on triodes and tetrodes by R.le Rossignol and his group.

'CVD' evolved from an informal arrangement between Signal School and GEC for the supply of valves under the USWA contract. In late 1938, when he was visiting le Rossignol, Coales was shown a small transmitter triode which in his opinion could be modified to provide a higher power replacement for the Marconi Osram DET12, which was giving a few kilowatts peak output in tests at Signal School. He could not discuss this with le Rossignol because neither he nor any other valve engineer had been cleared for access to information on RDF.

This difficulty was resolved almost immediately when C.S.Wright, Director of Scientific Research, Admiralty, gave his personal approval for selected individuals at GEC to be told about RDF. Within a few weeks, this was confirmed officially and a contract placed with GEC for 'Research and Development of Special Valves for Secret Applications'. For reasons of top security no reference was made to RDF, and to minimise the probability that the true purpose of the contract could be identified by those not party to it, it was decided to identify it as an addition to preceding Admiralty USW contracts by giving it the title 'Ultra Short Wave Communications Valve Development'. It soon became known as 'CVD'. Admiralty was by that time responsible for the research and development of valves for defence purposes, and one purpose of the contract was 'to enable DSR to ensure that valve development was suitably directed and initiated to meet the needs of all three Services'. Orders for new developments would be placed by DSR following recommendations from Signal School.

1.2 CVD and GEC: A Working Partnership

The first 'CVD' meeting was held at Wembley on 1st February, 1939, and attended by representatives of Signal School, including the Chief Scientist, G.Shearing, (Chairman) and H.G.Hughes, (Secretary), and by senior members of the Research Laboratories, including the Director, C.C.Paterson. It was agreed that protection of the commercial interests of the M.O.Valve Company would be dealt with on an individual basis as and when the need arose.

There was a review of developments in progress, including the E1011 'micropup' and the E1001 'acorn' triode, both designed to operate at 200 MHz. This was followed by a discussion of the possibility of making NT60 valves in a non-silica envelope to simplify manufacturing technology, and of the use of thoriated tungsten filaments to increase electron emission and reduce cathode heating power .

> Because the transmitter valves in radar systems are pulsed for periods up to 10 microseconds with a repetition rate up to 2000 pulses per second, anode dissipation is a fraction of that under continuous operation, even though the peak input power is many times that with DC input. Thus a reduction of cathode heating power can result in considerable easement of cooling requirements. Thoriated filaments were being used in broadcast transmitting valves, but it was thought that there could be problems with life and damage to the cathode surface arising from operation with anode voltages higher than those in use at the time.

At the second CVD meeting on 14th February 1939, the Signal School representatives were joined by Captain Willett RN, Captain Superintendent, and by R.A.Watson-Watt, Director of Communications Development, Air Ministry. (He was previously the first Director of the Bawdsey Research Station).

In addition to a complete review of the progress of development of valves for RDF and communications, priorities were allocated as follows:
(a) Valves for RDF.
(b) 'Micropup' transmitting valve E1011 for use at 200 MHz, and concentric version of DET12 for use in the band 300 MHz to 1 GHz (100 to 30 cm).
(c) Economy of cathode heating power.

(d) Research to obtain peak powers greater than 10 kW in the range 300 to 600 MHz.

(e) Receiving valves for Ultra Short Waves (300 MHz to 1 GHz). Watson-Watt stressed the need for operation down to 30 cm (1 GHz).

The review included magnetrons intended for use in Ultra Short Wave communications and RDF equipment. Coales of Signal School and Megaw of GEC were of the opinion that 'unless some definitely new proposals for operation at shorter wavelengths were forthcoming, no further development was required'.

Thereafter, the agenda of CVD meetings followed the same pattern of reviewing the progress and priorities of development and the allocation of samples to the Service Establishments.

At the third CVD meeting in May 1939, reference was made to 'Velocity Modulation' and to the publication in US journals of papers on the 'Klystron', including a suggestion from Stanford University that a single resonator could be used in oscillators. Megaw reported on split-anode magnetrons operating down to 20 cm. The Bawdsey R.S. representatives said that because of their existing commitments they could not start work at 30 cm for some time, but operation at this wavelength would definitely be required for systems in aircraft. Signal School did not wish to see work at 600 MHz prejudiced by work at other frequencies.

At the fourth CVD meeting held at Signal School in August 1939, GEC reported that they had successfully tested four samples of E1024, a metal-glass replacement for the silica envelope NT60 transmitting valve used in the coastal defence network and ground control interception systems. The new valve, later Type Approved as VT114, had a thoriated tungsten filament. In response to a request from F. Brundrett of Admiralty for early production to ease the pressure on Signal School for supplies of NT60, GEC offered deliveries in April/May 1940.

Megaw described the development of an 8-segment interdigital magnetron by H. Gutton of SFR in Paris. This had given ten watts CW at 16 cm, and been pulsed to 50 W in a simple RDF system which had detected returns from ships at ten kilometres.

On 12th September 1939, the first of a series of technical liaison meetings was held at Wembley, chaired by C.C. Paterson, and attended by R.A. Watson-Watt, DCD Air Ministry, members of the Laboratory, and representatives of Signal School. Paterson described the arrangements which had been made for the Laboratory to undertake pre-production of valves for the Services. There were plans to manufacture E1024 at 100 per week, beginning in mid-October. GEC were making cathode ray tubes for Bawdsey: CH and aircraft displays, a monitor tube for Signal School, and a copy of a Cossor tube for GL.

At the fifth CVD meeting on 29th September 1939, there was a discussion on standardisation between the Services of their requirements for cathode ray tubes. There was also some discussion on USW receiving tubes: British manufacturers were now making replacements for the RCA 'acorn' tubes; Dunham of GEC reported that the tetrode ZA1 was working down to 70 cm.

The meeting was followed by a meeting of the Cathode Ray Tube Sub-Committee, at which GEC outlined their programme of research, development and supply of tubes for Service requirements. W.B. Lewis of AMRE (Air Ministry Research Establishment, the new title of Bawdsey R.S., which had been dispersed from that location to Dundee and inland sites) commented on the use of 'modulation' methods instead of deflection in displays; intensity modulation of the spot gave a better response from small echoes but required shorter afterglow. [The reference to afterglow is indirectly a statement of the need for a faster build-up than that available from the double layer phosphors used in contemporary displays.]

The sixth and last of this series of 'CVD' meetings was held on 1st December 1939. In addition to the customary and lengthy review of progress and priorities of developments and allocation of samples, J.W. Ryde of GEC reported on the assembly and preliminary testing of velocity modulation tubes.

[Note: A Signal School group under R.W. Sutton was at this time making and testing 2-cavity klystrons based on information published by R.H. and S.F. Varian in the United States]

1.3 A Plan for the Future

On 14th November 1939, the 'Swinton' committee (the Sub-committee on Air Defence Research of the Committe of Imperial Defence) asked that 'The Board of Admiralty be requested to sanction substantial enlargement of the Valve Section of the Signal School for the purpose of future research and development to meet the needs of all three Services arising from the application of RDF'. The minute made no reference to a corresponding increase in production capability, even though Signal School were fully stretched in meeting demands for silica transmitting valves used in the coastal defence network and in Naval radar and communications. It would have been obvious to those directly concerned that the most effective, if not the only means for meeting the Swinton committee's request was to use expertise and facilities available in industry and the Universities but excluded from the RDF programme for reasons of security. Thus C.S. Wright, DSR Admiralty, who was responsible for research and development at Signal School and for the CVD contract at GEC, could justify formally and technically that other organisations (e.g. British Thomson Houston, Electrical and Musical Industries and Standard Telephones and Cables) be brought into CVD, and that the work at the Clarendon Laboratory, Oxford and by Oliphant's group at Birmingham be coordinated within CVD. This in turn would require the establishment of a new management structure for CVD.

On 16th November 1939, there was a discussion between

Director of Scientific Research, Admiralty
Director of Communications Development, Air Ministry
Deputy Director of Scientific Research, Ministry of Supply
F. Brundrett, SRE, Admiralty

DSR's note calling the meeting refers to development work being done for the Services by GEC, STC, Metropolitan Vickers, and J.H. Owen Harries. [Owen Harries was a small independent maker of special purpose valves. His attempts to develop devices for defence applications were ultimately unsuccessful, and so no further mention will be made of his participation in the CVD programme]

The note also refers to development of cathode ray tubes and oscilloscopes by GEC, Cossor and Baird. It goes on to say

'existing CVD meetings at Signal School and GEC have become somewhat embarrassing: not only do they last a long time, but they are attended by representatives of GEC, and in view of the number of other commercial organisations working on related problems, it is becoming difficult to discuss such questions openly since such action would probably place the GEC in a privileged position.

There is also some hesitation in arriving at a conclusion that the apparatus produced by another firm is more suitable than that produced by GEC. It is obviously undesirable that policy discussions take place in the presence of firms' representatives.'

After a short discussion it was agreed that 'future CVD meetings be held in two parts. The first should be held in Admiralty, and attendance limited as far as possible to one administrative and one technical representative from each Service in addition to the secretary. Technical discussion at this meeting should be restricted to that required to decide policy.'

'The second part of the meeting should be held at the research organisation of one of the firms engaged on development work (on valves). While at present it was probably desirable to continue meeting at GEC, the possibility of meeting at laboratories of one of the other firms should not be neglected. Attendance would be as desired by Establishments. The meetings would be primarily for instructing the firm on the priority in which work be continued, and for purely technical discussion.'

The first Valve Development Policy Committee meeting was held on 30th November 1939. Those present were:-

C.S. Wright, Admiralty (Chairman)
F. Brundrett, SRE, Admiralty
G. Shearing, Chief Scientist, Signal School
W. Ure, Signal School
H.G. Hughes (Secretary), Signal School
DDSR, Ministry of Supply
D.H. Black, ADEE*
L.S. Harley, DDCD, Air Ministry
W.B. Lewis, AMRE

[*the War Department Air Defence Experimental Establishment at Christchurch, which included members of the team previously at Bawdsey]

DSR opened by saying that the meeting had been called as a result of the discussion on 16th November. 'Whereas only GEC had been brought into the work (on valves for RDF) it now affected Metropolitan Vickers, ST & C, Birmingham University, and the Clarendon Laboratory, Oxford. The purpose of the discussion had been to improve coordination of valve development for the Services. This was DSR's responsibility, and its scope was extending rapidly. It was now evident that discussions of policy and technical progress should be separated. Policy was for discussion by the Services alone. Arrangements had to be made for safeguarding commercial security of the various companies.'

It was agreed that 'policy matters be discussed at meetings of the committee now assembled, meeting at Admiralty. Technical matters would be discussed at a separate Committee of representatives of the Service Establishments only, meeting every two months at one of them. Signal School would be the most convenient. Meetings previously held at Wembley would be discontinued, and visits to the various firms arranged as and when considered necessary by any of the Establishments, all Services being informed'.

Monthly progress reports would be prepared by

    GEC on the work at Wembley

    Ministry of Supply on the work at STC

    Air Ministry on Cathode Ray Tubes

DSR would keep the Committee informed of the work at Birmingham and the Clarendon Laboratory.

It was also agreed that an offer by the Sperry Company of America to bring klystrons to England be investigated, and that Standard Telephones and Cables Ltd (STC) be asked to make the arrangements.

1.4. From Partnership to Participation

The Policy Committee meeting on 30th November 1939 was the beginning of a new phase in the development of valves for the Services. Although GEC was by far the largest contractor,the structure of CVD now envisaged participation on equal terms by other companies as the requirements of the Establishments diversified and increased. CVD had become a system for

Coordination of Valve Development

The acronym remained appropriate and survived through the war and for many years after.

The first Secretary of CVD, who was responsible for technical administration and acted as Secretary of the Policy and Technical Committees, was Professor R.A. Whiddington F.R.S., seconded from Leeds University in May 1940. In 1942, he was joined by J.H.E. Griffiths, who was at that time head of the microwave group at the Clarendon Laboratory, and in 1943 succeeded him as Secretary of CVD. In late 1943, W.J. Picken of the Marconi Company was attached to CVD. Known with respect and affection as 'Uncle' he took over from Griffiths, who returned to Oxford at the end of the war.

The first CVD Valve Development Technical Meeting was held at Signal School on 16th January 1940. It was attended by G. Shearing, Chief Scientist, Signal School (Chairman), H.G. Hughes (Secretary), C.E. Horton, R.W. Sutton, A.A. Symonds, J.F. Coales, H.M. Bristow, and B. Hodgson all from Signal School; L.S. Harley, Air Ministry. W.B. Lewis, AMRE; D.H.Black, P.E. Pollard and C.W. Oatley, all from ADEE [GEC had circulated previously a report on their work in the period February to December 1939,(Appendix 1).]

In addition to a review of progress in development and discussion of new requirements, it was reported that STC had designed and made a replacement for NT57T used in CH transmitters, and agreed that Standard Telephones and Cables Ltd be recommended to CVD Policy Committee as a second source for valves designed by GEC. [For many years before 1939 STC and MOV had been the largest British makers of valves for broadcasting and communications transmitters.]

AMRE tabled a new requirement – a valve giving 200 kW peak at 200 MHz which had to be a direct replacement for valves used in existing equipment. This was supported by Signal School. Because of the existing load of work on GEC, placing of the new requirement was referred to Policy Committee.

Following a proposal by Black and Horton that fuller use be made of research facilities other than those at GEC, it was agreed that Policy Committee be asked to consider this, noting that this would require more workers knowing about RDF.

Air Ministry had placed a contract with GEC for the development of Cathode Ray Tubes. This was similar in format to the CVD contract. Reports on progress would be sent to the other Services. Signal School asked that the contract cover requirements for the Navy in the same way as Navy developments were made available to the other Services.

W.B. Lewis had tested samples of EFF50, a double tetrode supplied by Mullard Philips, which operated up to 300 MHz.

At the liaison meeting in January 1940, GEC stated that they had been asked to develop a complete AI system working below 50 cm, including the transmitter and receiver. Although the team was outside the CVD group, valve developments would be reported to the Technical Committee. Signal School stressed the need for more research on receivers in this frequency band. There was also a need for a low noise tetrode on an EF50 base to work at 90 MHz.

The second Policy Meeting was held on 24th January 1940, attended by C.S. Wright, Director of Scientific Research, F. Brundrett and H.G. Hughes of Admiralty, L.S. Harley of Air Ministry, and J. Cockcroft, Director, and D.H. Black of ADEE. Black explained the Technical Committee's proposal for participation by firms other than GEC. It was agreed:-

(a) that a Technical Committee be set up with STC in a similar way to that at GEC.

(b) that Harley survey and prepare a report on available research facilities.

Hughes and Harley stressed the need for planning of valve production. There was also a need for a Valve Life Test Laboratory which would relieve the load on test facilities at development laboratories and permit extended life tests to be made.

It was agreed that GEC and STC be asked to work on the requirement for 200 kW peak at 200 MHz. Signal School would attempt a silica version as soon as their other commitments made this possible.

F. Brundrett questioned the advisability of having so many organisations working on 'klystron' type apparatus. The meeting supported the opinion of Black and Cockcroft that as many people as possible be involved in the new principle.

At the liaison meeting in February 1940, J.W. Ryde of GEC reported that he was continuing his theoretical studies of velocity modulation; an experimental receiving valve had been assembled and DC tested, but no functional tests had been made.

The Second Technical Meeting was held on 15th March 1940. The committee was 'gravely concerned with the inadequate supply of valves suitable for reception at and below 1.5 m. Acorns are the only proven valves in production. Present manufacturing capacity is totally inadequate'. [At the January liaison meeting, GEC had said that a major difficulty with acorn fabrication and assembly was that the girls were unable to continue working for long periods, and that training of new recruits took several weeks].

At the third Policy meeting on 27 March 1940, it was agreed that 'full coordination of all research and experimental development work on valves for all purposes for the Services was most desirable'.

The committee recommended that 'DSR Admiralty be responsible for such coordination, working through the CVD Policy Committee.'

It was agreed in principle that 'it is most desirable that one man be responsible for valve production and supply'. Admiralty and Ministry of Supply were agreeable to his being located in the Ministry of Supply. Air Ministry deferred their approval of this arrangement.

DSR said that he had asked Birmingham University and the Clarendon Laboratory to work at 5, 10 and 50 cm so that results obtained on transmitting and receiving devices would be complementary. He added that GEC had been instructed to develop an AI system on 10 cm. [Some years before, E.G.Bowen of Bawdsey and DSR had agreed that AI would ultimately require operation at 10 cm. Even so, several thousands of 200 MHz AI and ASV sets were built and used throughout the war.]

It was agreed that DSR act on the proposal for a Valve Life Testing Laboratory. H.G. Hughes of Signal School had located a suitable site.

It was agreed that because of their connections with Philips in Eindhoven in occupied Holland, Mullard could not be given secret work, but maximum use should be made of their research and development facilities for 'non-secret work'.

At the GEC liaison meeting in April 1940, J.W. Ryde discussed his theoretical work on klystrons. Signal School reported that they had obtained 100 W from a demountable unit. A large part of the time of the meeting was occupied by discussion of the progress of developments, even though a progress report had been circulated previously.

At the third Technical Meeting on 17 May 1940, reference was made to a meeting between Policy Committee and STC, who had offered to develop a valve giving 200 kW peak at 200 MHz (to meet the AMRE requirement) and a replacement for the RCA 'Acorn' to operate at 600 MHz. During a discussion of how the use of Mullard facilities for research and development would be affected by their loss of contact with Eindhoven, it was reported that Philips had found that crystals were as good as the best diodes they could make for centimetre wave reception.

[It is almost certain that 'centimetre wave reception' refers to wavelengths greater than 20 cm i.e. above the microwave band. At the time of the meeting, GEC had made and tested tungsten-silicon catswhisker crystals in their experimental 25 cm AI apparatus. It is not outside the bounds of probability that the recorder of the meeting misheard or was mistaken in his identification of the source of the information.]

AMRE and ADEE stressed the importance of finding a firm to develop manufacture of crystals on a small scale. [H.W.B. Skinner of AMRE was at the meeting. By July 1940, he had made and tested tungsten-silicon 'cats-whisker' crystal mixers (see 4.4. 'Microwave Crystals') in his experimental radiolocation system.]

AMRE said they wished to use the same aerial for transmission and reception at 10 cm (3 GHz), (see 4.5. 'TR and TB Cells').

Development of velocity modulation valves at the Clarendon Laboratory was to be concentrated on single resonator types. Their design envisaged the use of secondary emission from the 'collector' to generate the return beam. Signal School were continuing their work on power klystrons and were designing a low power oscillator with a reflector electrode.

It was reported that Megaw of GEC had visited Randall and Boot at Birmingham University to discuss their demountable 10 cm magnetron. Following a detailed discussion of their results, he was designing a sealed-off unit capable of being used in a conventional electromagnet. It would incorporate a block from Birmingham. (A more detailed account of this early work and the subsequent history of development of radar magnetrons is given in 4.2. 'Magnetrons').

At the fourth Policy Meeting held on 27 May 1940, DSR reported that M of S concurred with the proposal that DSR Admiralty, working through the CVD Policy Committee, be fully responsible for 'coordination of research and development of valves for the Services'. No reply had been received from Air Ministry. L.S. Harley stated that they would concur. He would press for official confirmation.

DSR reported that he would be meeting representatives of GEC, STC and Mullard, saying that it was highly probable that they would be fully prepared to cooperate within the CVD system. It was agreed that British Thomson Houston be brought in at a later date.

Air Ministry did not agree with the proposal for 'one man management' of valve production for the Services. It was therefore agreed that a committee be set up to coordinate requirements before sending proposals to the separate contracts departments.

At the GEC liaison meeting in May 1940, Megaw discussed his design for a pulsed 10 cm magnetron based on the Randall and Boot anode block, and tabled a sample of an 8-segment interdigital magnetron with a large oxide cathode made by H. Gutton of SFR in Paris. One of the GEC multicavity models (E1189) would be fitted with a similar cathode.

Signal School Bristol reported that they had obtained 200 mW output from a 10 cm reflector klystron. Ryde of GEC was continuing his theoretical work on rhumbatron resonators and had concluded that operation at a few hundred volts would require grids to be fitted to the horns of the resonator to improve efficiency and simplify design of the electron beam system.

The fifth CVD Policy Committee Meeting was held on 22 July 1940. It was reported that Air Ministry concurred officially with DSR's proposal re CVD. It was agreed that Electrical and Musical Industries Ltd and BTH be brought into CVD.

The Interservice Valve Production Committee was in operation.

1.5 Evolution for Efficiency; Delegation and Specialisation

Soon after the Policy Committee meeting in March 1940, it became obvious that the CVD system had to expand if it were to fulfil its technical function efficiently. Because of its small number of members, the Technical Committee by itself would be unable to review the progress and assess the merit of an increasing number and variety of developments at contractors' laboratories. The number of developments in progress, and requests for new ones, would be too large for adequate assessment of their merits within the time and technical expertise that members could make available.

To overcome these difficulties, the activities of the Technical Committee were supported by a growing number of auxiliary and sub-committees. For example, the cathode ray tube committee which began ad hoc as part of the original series of meetings with GEC, was attached formally to the Technical Committee, and given responsibility for progressing and review of all cathode ray tube development for the Services. The Committee for Cathode Ray Tube Development, very soon known as 'CCRTD', was also required to recommend action by Technical Committee and, through them, by Policy Committee. In September 1940, the Sub-Committee on 'Modulation Techniques' had its first meeting. Its general terms of reference were similar to those of CCRTD. In addition to the committees reviewing research and development, others were set up with more specific objectives, for example the '50cm' and '10cm' committees, which were asked to report on the state of the art in those wavelength bands and to recommend action to speed up research and development.

Delegation by Technical Committee was a continuing process, with changes in the number, size and membership of the sub-committees reflecting the needs of the times. For example, the Sub-Committee for 'RDF Valve Development' had its first meeting on 12 October 1942, with over fifty members at the morning session attended by contractors' and Service representatives in roughly equal numbers. The agenda included items previously covered in detail by the Technical Committee and '50cm' and '10cm' Committees. The Committee on Modulation Techniques and CCRTD were similar in numbers and representation.

At first sight, the size of the review committees would suggest that they could not operate effectively, but they were certainly the most efficient means available at the time for discussion between specialists in the subjects under review and for communication between all those concerned with a particular development, including the administrators. [Their chairmen, or their representatives if chairmen were not available at the time, attended meetings of the Technical Committee.]

This continuing process of evolution made CVD an efficient system of project management that survived for many years. It had an hierarchy, but this was part of the machinery of operation rather than a format for administration. The strength of CVD was the individual knowledge and authority of its members, irrespective of seniority. Apart from matters of policy, the history of CVD is the history of the development of the active devices they produced, which is outlined in Part 2.

1.6 CVD Policy and Technical Committees

The 6th Policy Committee meeting was held in September 1940. It was agreed that research on reception in the 50 cm band be continued because Army and Navy equipments would continue to operate around 50 cm. Development of measuring apparatus would be coordinated by DCD Air Ministry.

It was reported that the Signal School group under R.W. Sutton at Bristol had made a sealed off 10 cm reflex klystron with 3% tuning (the first 'Sutton tube').

At the Policy Committee meeting in October 1940 there was discussion of a suggestion emanating from the RDF Committee that ADEE and TRE* might abandon work at 50 cm to speed up development of S-band (10 cm) equipment. This was accepted; Signal School would continue to work at 50 cm because of their commitment to Naval fire control systems. [Note: A decision would soon be made by ADEE to use 10 cm high power magnetrons in the GL3 gun-laying equipment. TRE were developing 10 cm AI based on the GEC S- band magnetron E1198.]

[*Following the moves to Swanage of the dispersed groups of AMRE, the new location was for security reasons described as 'TRE'. It was later known as the Telecommunications Research Establishment, which became part of the Royal Signals and Radar Establishment (RSRE) at Malvern.]

The demand for S-band (10 cm) reflex klystrons from Bristol was now up to 60 samples. It was agreed that E.K. Cole be given an order for 100, to include the cost of tools.

[Before 1939, E.K. Cole had manufactured receiving tubes for their own broadcast receivers. R.W. Sutton was responsible for their development. He joined Signal School after the operation was bought by Mullard. The facilities remained in charge of A.R.Chance, who had worked with Sutton for several years previously, some with Ferranti.]

It was agreed that 'there was a need for some central authority to reduce the number of types of valve used by the Services, and that if this came about, the CVD Production Sub-Committee should cease to exist'. [Note: This foresees the formation of ISTVC, the Inter-Services Technical Valve Committee, which was responsible for issuing the agreed CV specification and for deciding which valves be placed on the preferred list (see 7.6).]

The GEC contract had been extended. A contract with STC was under discussion. BTH were considering whether they would come into the CVD system. DSR had asked GEC to contact BTH with a view to their manufacturing valves for RDF, using GEC designs. GEC and MOV had agreed to the disclosure of information on development. [BTH were already engaged in development and pre-production of magnetrons and thyratrons as part of their GL3 contract with M of S, working with the Birmingham University group under M.L. Oliphant.] It was reported that EMI were making valves for experimental S-band equipment as part of a contract with the Ministry of Aircraft Production.

At the Technical Committee meeting in November 1940 it was stressed that work on crystal mixers was receiving inadequate attention.

The Policy meeting in December 1940 decided that BTH would be used for specific developments only, and that no further action be taken on a general contract. [At the time of the meeting T.H. Kinman of BTH was supplying samples of S-band crystals to Signal School, TRE and the Birmingham group. These were based on H.W.B. Skinner's 'cats whisker', using a ceramic capsule assembly to obtain adequate mechanical stability. BTH were also making S-band magnetrons to GEC designs for airborne and Naval radar.]

At Technical Committee meetings in January 1941, it was reported that Signal School, GEC and EMI were making S-band reflex klystrons. The supply of magnetrons from GEC was inadequate, and it was recommended to Policy Committee that STC Ilminster be asked to consider manufacture to GEC designs.

At the GEC liaison meeting in January 1941 it was stated that development of 3 cm (X-band) magnetrons had priority over 10 cm types because E1189 and E1198 were going into production. Oliphant's group at Birmingham University had made a working sample of a 3 cm tube. GEC were asked to send 10 cm magnetrons to USA as soon as possible. They asked for samples of Western Electric types in return. [Note: A sample of E1189 was taken to the USA by the Tizard mission in August 1940].

At the Policy Committee in January 1941, it was agreed that STC be asked to second source GEC. [Note: STC did not undertake to manufacture magnetrons. Production capacity was increased by BTH building a production unit near Rugby and later by Marconi's Wireless Telegraph Valve Co. (now English Electric Valve Co.) coming in as a second source to MOV].

EMI had a contract from the Ministry of Aircraft Production for the development of X-band (3 cm) measurement techniques and apparatus which could involve making valves. It was agreed that the placing of a CVD contract be deferred until it was known which wavelength would be used operationally below 10 cm – there might be a preference for 5 cm instead of 3 cm.

At the meeting of the Technical Committee on 6 February 1941 it was agreed that:-

'the Technical Committee should not discuss items on the agenda of Sub-Committees unless attention is called to a particular item by the Sub-Committee or a member of the Technical Committee'.

'Policy Committee be asked to hasten incorporation of the W/T Board Technical Sub-Committee for Coordination of Valve Development into the CVD organisation with the new title 'Communications Valves Sub-Committee'.

'GEC liaison meetings be held at the GEC Laboratories' [Note: these were previously held at Signal School. The change of venue made it possible for members of Technical Committee to see the work in progress. It became standard practice to hold liaison meetings at contractors' laboratories.]

At the meeting of Policy Committee in February 1941, DSR commented on 'the tendency to deal too much with detail and to review too closely the reports of the Technical Committee and Sub-Committees'. The W/T Board had agreed to 'the formation of a Communications Valves Sub-Committee reporting to the main CVD Committee'

A review of Sub-Committee activities would be made to minimise overlap in their agenda.

At the 10th meeting of the CVD Technical Committee held on 18 March 1941 it was agreed that it be attended by chairmen of Sub-Committees or their representatives. Valves for action by the Inter-Service Technical Valve Committee (TVC), which had been in operation for about nine months, would be a standard item on the agenda. It was agreed that 'velocity modulation tubes' (reflex klystrons) be recommended for action by TVC. [TVC was chaired by

C.C. Paterson of GEC, who referred to it in jest as the 'Anti-Valve Committee' because one of its important functions was to minimise the number of different types of valve accepted for use by the Services, and attended by representatives of the design Establishments and their headquarters Directorates. The Secretary was H.G.Hughes of Signal School. It was provided with technical and secretarial assistance by the Ministry of Aircraft Production. Subsequent to recommendations by CVD Technical Committee, it was responsible for formal Type Approval of valves developed for Service applications, for issue of the agreed CV specification prior to production in quantity (in cases of urgency, deliveries could be made to a provisional specification), and for selection of valves to be placed on the Preferred List. MAP also provided secretarial and technical assistance to the Interservice Valve Production Committee, chaired by Air Commodore Leedham of Air Ministry and attended by representatives of the Ministry of Supply, Admiralty, and the Board of Trade (covering civil interests). The Secretary was D.N. Truscott of MAP, who early in 1940 had initiated action on a preferred list of valves to be used in Air Ministry equipment, and was formally responsible for the periodic revision and re-issue of the common Preferred List for all three Services (see 7.6 and Appendix 2).]

It was agreed that magnetrons have higher priority than klystrons at Birmingham University. Reference was made to discussions with BTH on 'making magnetrons to GEC instructions' for use in GL3 equipment being made at Rugby. It was suggested that BTH also make tubes for Birmingham. [Note: By this time the decision to use magnetrons and reflex klystrons in GL3 was firm. In December 1940 there had been discussions between BTH and GEC regarding the supply of magnetrons giving at least 100 kW peak output. In the months that followed, close working arrangements grew up between the teams at Wembley and Rugby, with GEC concentrating on valves for airborne applications and BTH on high power types for ground based equipment. BTH also worked for a short time on reflex klystrons as an alternative to the NR89 developed at Bristol by the Signal School group, and followed this by research and development on 'all-metal' X-band reflex klystrons].

At the CVD Technical Committee in July 1941, it was reported that the third ISTVC list of Preferred Valves for Service Use was about to be published. [In November 1941, ISTVC issued a complete Service list of Valves for Special Purposes (see Appendix 2).]

The Valve Life Test Laboratory would be working before the end of the month.

It was reported that Policy Committee had reduced the priority of power klystron work at BTH. The '10cm' Valve Committee had recommended that, because of their work load, GEC should restrict their plans for work on reflex klystrons to the making of 500 NR89s, and that orders be placed on E.K. Cole for production of NR89s and other S-band tubes required by the Services. [Soon after this recommendation was made, GEC were requested to cease work on reflex klystrons so that all their available effort could be concentrated on other microwave devices. (Between March and December 1941, GEC developed an equivalent of NR89 and supplied about sixty samples in all to Signal School, TRE and ADRDE. They stopped work on klystrons in early 1942; Ryde then concentrated on the metallurgy of microwave crystals)]. A CVD contract had been placed with EMI, who had developed an S-band reflex klystron giving 300 mW output. Policy was not yet determined for a group at Oxford working on centimetre valves. [At this time, the Clarendon Laboratory was testing 3 cm TR cells, local oscillators and crystals. Before the end of 1941, they were developing 3 cm reflex klystrons in collaboration with EMI and Sutton's group at Bristol. Later on they became the accepted British authority on microwave receiver measurements, and concentrated their valve development on extending the range of operation of reflex klystrons and magnetrons down to 1.25 cm.]

At the time of writing, no records of CVD Policy and Technical Committee meetings after July 1941 have been located. By that time the sub-committee system was well established, and supervising research and development of devices within the terms of reference of each of the specialist subject committees. Because the activities of the Policy and Technical Committees, including the choice of projects and the relative priorities allocated thereto, were reflected in the agenda of the Sub-Committees, the history of CVD thereafter is essentially that of the devices created by its members (see also Appendix 3).

Part 2

Active components for radar systems

Chapter 2

An Introduction to Radar Technology

2.1 From metres to microwaves

2.1.1 Foreword

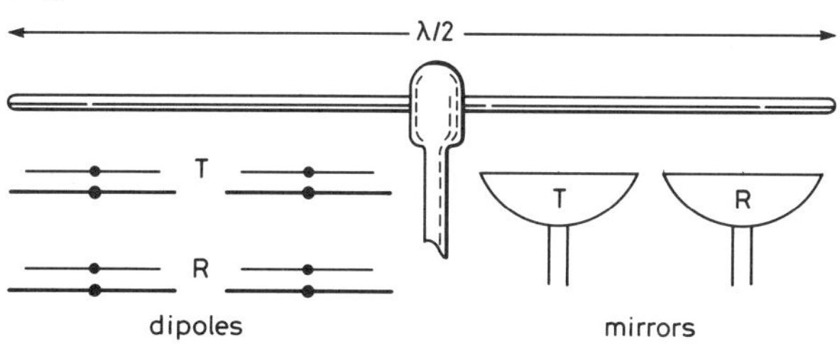

In simple terms, a radar system comprises a radio transmitter which illuminates a target area by means of a directional aerial system, a receiver, also attached to a directional aerial, which detects radiation reflected from objects within the target area; and a display which indicates their range or their bearing from the receiving aerial, and preferably both.

2.1.2 Range measurement

The simplest method for measuring range is to radiate a pulse of energy and then measure and display the time between transmission and the detection of a reflected pulse. For an object at 1500 metres, this is approximately 10 microseconds, and so the pulse duration must be one microsecond or less if objects close together are to be differentiated in range. When high accuracy of measurement is required, the radiated pulse must rise to its peak amplitude in a small fraction of one microsecond. Thus the device which switches on the transmitter must generate a high power high voltage pulse which rises to its peak value in a comparable and preferably shorter time. The rate of rise of the RF pulse is limited by the fundamental characteristics of resonant circuits in the transmitter (their 'Q's'), and so several cycles of oscillation are required for the radiated signal to reach its peak amplitude.

Some land based and shipborne installations used switches developed from industrial mercury vapour thyratrons. These were not suitable for operation with repetition rates approaching one thousand pulses per second because of the time taken for the vapour to deionise after the pulse had passed. They were also unsuitable for use in aircraft because of the fall in vapour pressure caused by the low ambient temperature at high operational altitudes. This problem was solved through development in 1942 of the 'trigatron', a cold cathode triggered spark gap filled with a mixture of argon and oxygen at a pressure of several atmospheres. An alternative solution for powers up to 100 kW peak was found through the development of hard valve modulator tubes. (see Chapter 6. 'Modulator Switches')

Only a very small fraction of the transmitter power reaches the receiver after being reflected from the target, and so the development and manufacture of radar systems depends equally upon the availability of transmitters giving high peak power output (at least tens of kilowatts) and broadband receivers giving a good signal to noise ratio* with inputs of a few microvolts at frequencies above 20 megaherz. It so happens that this is about the minimum frequency at which objects such as fighter aircraft are reasonably efficient reflectors of incident energy.

*The signal to noise ratio or noise factor of a receiver limits its ability to detect very small signals because it determines by how much the noise content of an incoming signal is increased by electrical noise originating in the front end of the receiver itself. As the signal decreases in amplitude it will ultimately be swamped by the 'front end' noise (see Appendix 4)

To measure ranges up to 150 km without ambiguity arising from the reception of energy from successive transmitter pulses, the interval between them must be greater than one millisecond. When maximum range of detection is a prime criterion of performance, pulse durations up to about ten microseconds can be used. This increases the amount of energy impinging on the target during a pulse. It also permits a reduction in bandwidth of the receiver, which increases its ability to detect returning signals in the presence of external interference or noise generated in the input stage of the receiver itself.

2.1.3 Bearing measurement

The bearing of an object can be measured by 'floodlighting' the target area by means of a transmitting aerial of low directivity, and using a conventional direction finding (DF) receiver to indicate its position.

When maximum range and accurate measurement of bearing is required, the most efficient use of transmitter power is obtained by using transmitting and receiving aerials with the same directivity, mounting them so that they are accurately aligned, and rotating them in synchronism to scan the target area. As the beam width is reduced to increase the accuracy of bearing measurements, there is an automatic increase in the maximum range of detection because more power is concentrated on the target and the receiver is less liable to interference from other sources. This gain is offset by the need to increase the rate of rotation of the aerial system to minimise the probability that a moving object will not be detected during its transit of the target area.

In general terms, the beam width of an aerial system is inversely proportional to the number of elements therein, and so to its dimensions along or perpendicular to the axis of the beam measured in terms of the wavelength of operation. [Arrays can be designed with a series of dipoles in either direction. Some domestic TV and VHF receivers use directional aerials comprising an array of 'halfwave elements' spaced along a horizontal support attached to a high point such as a chimney stack and pointing towards the nearest transmitter.]

The physical dimensions of an aerial system are limited by mechanical considerations such as windage in land-based arrays mounted on high masts to give the best possible coverage of the target area. Similar restrictions apply to rotating arrays for shipborne early warning radar, which are mounted at the masthead to obtain all-round viewing. In applications such as turret mounted fire control systems, mobile gun laying equipment and aircraft armament, the size of the aerial system is determined absolutely by the space available. Continuing demands for smaller size or improved resolution can be met only by operating at shorter and shorter wavelengths.

Thus each major improvement in equipment performance has to be preceded by the development of active devices operating at higher and higher frequencies, or generating higher power to increase range when the operating wavelength provides adequate resolution and receiver performance cannot be improved significantly.

As the wavelength is reduced, multiple element arrays become correspondingly more difficult to engineer. At frequencies above one gigaherz (the microwave band), the aerial is in the form of a horn attached to a waveguide, or more often a paraboloid mirror with a radiating element at its focus. A mirror one metre in diameter gives a beam width of approximately 6° at 10 cm wavelength (3 GHz) and 2° at 3 cm (10 GHz). The ultimate reduction in size is achieved when a single mirror is used for transmission and reception. In this 'common T and R' mode of operation, power from the transmitter must be prevented from entering and so damaging the receiver, and signal from the target be prevented from entering the transmitter and so reducing the effective sensitivity of the receiver. Common T and R systems were also used in metric radar.

In metric radar, extension of the frequency range up to 600 MHz was achieved by the development of new fabrication technology for triodes and tetrodes. The 'micropup' series of high power transmitter valves developed by GEC had short, comparatively large diameter cathodes to provide high peak emission during the pulse, 'parrot cage' grids attached to metal annuli sealed through the glass envelope, and external copper anodes. Because the valves are turned on for less than 20 microseconds every millisecond, high peak emission, rather than cooling of the grids and anode, is a prime criterion in design. Thus the overall size of the assembly is smaller than that of a corresponding unit designed for continuous operation. This, and the use of disc seals, reduced lead inductance and so reduced the minimum wavelength of operation. It also permitted the very close spacing of the electrodes which is essential to efficient operation at ultra short wavelengths. The signal amplifiers and low power oscillators developed by STC and GEC had disc seals for the planar grids and internal anode, and grid to cathode spacings of the order of one tenth of a millimetre. [An outline of the development of active devices for metric radar is given in Chapter 3]

Early Metric Radar. The simplest directional aerial is a 'doublet' comprising two half wave dipoles one behind the other, spaced one quarter wavelength apart in the same plane. It gives a beam width of roughly 40° between the half-power points in the plane of the dipoles and 90° in the perpendicular plane. Suspending a number of horizontal doublets one above the other reduces the vertical divergence, producing a concentrated 'floodlight' beam. Arrays of this kind were used as transmitting aerials in coastal early warning stations (CH). Operating at 26 MHz (13 M) they could be suspended safely between pairs of masts 300 feet high. A pair of horizontal doublets spaced vertically by one half wavelength gives a beam width around 50° in the vertical plane. Spacing them horizontally by one wavelength gives a beam width of roughly 20° in the horizontal plane. The first shipborne early warning system, Type 79, used separate transmitting and receiving aerials comprising two horizontal doublets one above the other. Operating at 43 MHz, they occupied a swept volume roughly 4 metres in diameter. Type 281, which followed Type 79, used four doublets to halve the horizontal beam width and increase the accuracy of bearing measurements. It operated at 90 MHz, and so the volume swept by the rotating array was no larger than that of Type 79. Type 284, the first Naval fire control system, operated at 600 MHz, and used transmitting and receiving arrays having twentyfour dipoles in horizontal cylindrical paraboloid mirrors to obtain a 3° beam in the horizontal plane. The mirrors limited vertical divergence to about 30°. The overall length of the array was about thirteen wavelengths (6.5 metres). The first operational airborne radar systems (Air to Surface Vessel and Air Interception) operated at 200 MHz (1.5 m). At this frequency, an array of four horizontal dipoles for transmission could be mounted fore and aft on top of the fuselage of an ASV patrol aircraft, and one array along each side for reception. This gave beam widths of 20° to port and starboard for location of targets as the area was swept by traverse of the aircraft. For homing on a target, ASV was fitted with forward looking aerials mounted on the nose. AI aircraft were fitted with arrays on the nose and the wings.

2.1.4 Microwaves – the new technology

From late 1940, when the first high power magnetrons and reflex klystron local oscillators were available for development of equipment, operation in the microwave S-(10 cm) band provided a major extension to radar technology. It could be said that microwaves changed the whole concept of design, particularly in mobile and high definition systems. For the first time, beam widths of a few degrees could be obtained from aerial arrays which were small enough to be mounted on vehicles and gun platforms, or housed within the nose or in streamlined enclosures on the wings or fuselage of an aircraft. Operation at microwave frequencies also increases the probability of detection of small objects such as periscopes or marker buoys, which become efficient reflectors at very short wavelengths (see also 2.1.1).

> Early Microwave Radar. Type 271, the first Naval S-band radar, was in use by the end of 1941. Designed specifically for installation in escort vessels, it transmitted powers up to 80 kW peak and used two cylindrical paraboloid mirrors ('cheeses') stacked one above the other, giving a beam width of approximately 2° in azimuth and 20° vertically. Its great advantage over systems operating below 1 GHz was its ability to detect small targets at sea level, such as the schnorkel of submarines. GL3, the first Army microwave radar was operational early in 1942. It used separate 1.5 m mirrors on opposite sides of the gun mounting for transmission and reception, and transmitted powers up to 500 kW at 10 cm. The first airborne microwave radar systems were AIMkVII and H_2S, which operated in S-band with powers up to 40 kW peak and used a single paraboloid mirror for transmission and reception. H_2S was designed to provide terrain recognition for navigators of night bomber aircraft, using a downward and forward looking rotating beam to distinguish between open country and 'hard' targets such as buildings, railway lines, bridges and other structures. The return signals were displayed on a Plan Position Indicator (see 5.3) so that a comparison could be made between the picture on the screen and maps of the locality. The early equipments, which were in use by Autumn 1942, gave inadequate definition over large built up areas, and so the wavelength of operation was changed to 3 cm (X-band) in 1943.

The development of microwave radar was made possible only by the availability of an entirely new generation of power generators. In the triodes and tetrodes used at ultra short and longer wavelengths, power gain is obtained by amplitude modulation, i.e. by using a grid placed close to the cathode to vary the number of electrons leaving the grid-cathode interspace and so the amount of current reaching the anode. Because electrons leaving the cathode take a finite time to pass through the grid to the anode, increasing the frequency of the signal applied to the grid will result in a condition where electrons leaving the cathode during a positive half cycle will overtake those which left during the previous negative half cycle, and so reduce and finally destroy the amplitude modulation of the anode current. This 'transit time' effect can be reduced by very close spacing of the electrodes, but this remedy is limited by the difficulty of maintaining close tolerances in structures subject to thermal or other mechanical stress. Ultimately, the transit time effect can be eliminated only by designing active devices in which the effect of variation in velocity of the electrons is itself used to obtain power gain. This is the principle of 'velocity modulation', upon which the operation of all microwave power generators is based. [An outline of the history of development of critical active components in microwave radar systems is given in Chapter 4.]

2.1.5 Displays
When a radar is directed towards a target to measure its range, the information is displayed on an A-scan, which uses a cathode ray tube similar to those in oscilloscopes for the display of waveforms in communications equipment. A trigger signal from the device which switches on the transmitter causes the spot to traverse the screen from left to right during one pulse interval. The signal received from the target is displayed as a vertical spike at a displacement corresponding to its range. Because the A-scan screen is re-energised several hundred times per second, there is no flicker. When a narrow beam radar is used for 'all round' viewing, a Plan Position Indicator is used. Here the blacked out spot moves radially from the centre of the tube and rotates in synchronism with the array. The received signal generates a bright spot at a position corresponding to the range and bearing of the target. Displays of this type were first proposed by E.G. Bowen of Orfordness in 1935. In a PPI, the screen of the cathode ray tube must retain the images of targets for more than one rotation of the array, so that the observer has before him at all times a useful picture of the target area. It is an advantage if the display can be viewed in ambient light sufficient for the operator to perform other tasks as required. Thus the phosphors used in the tubes must give adequate brightness from the afterglow which persists when the trace has passed, and retain this image for long enough to provide an acceptable picture, but not so long as to cause masking of new images by those recorded earlier, for example when the 'heading' of the display is changed as a result of a change of course by the ship or other vehicle in which it is installed. (see Chapter 5 'Cathode Ray Tubes for Radar Displays')

2.2 Basic Radar Systems

2.2.1 Metric Systems (up to 600 MHz)
In the twin array metric system shown at (a), the aerials T and R, which may be remote from the rest of the installation, are connected to the transmitter TX and the receiver RX by polythene insulated coaxial cables. [The transmitters

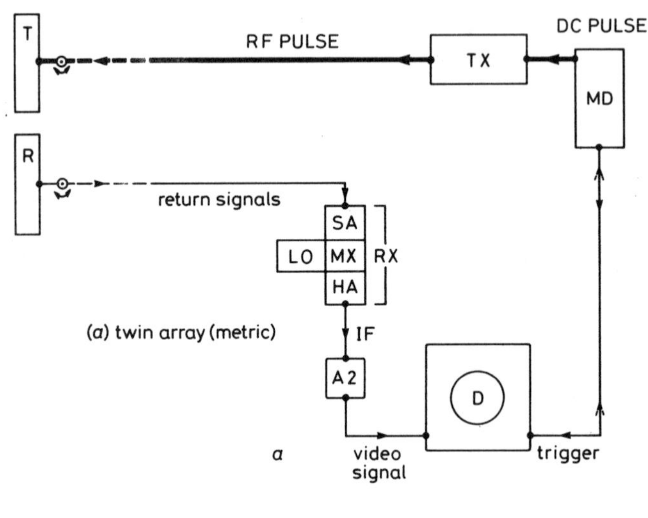

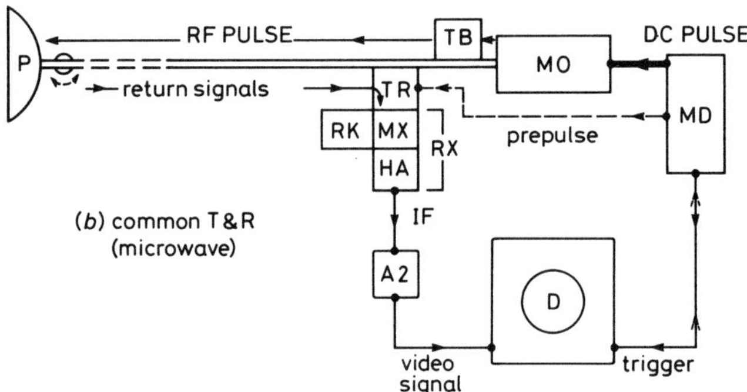

used a variety of circuit configurations based upon a self-oscillator which defined the frequency of the radiated signal, sometimes with one or two stages of power amplification between it and the aerial system.] Because of the rapid rate of rise and the high peak current in the pulse, the transmitter and the modulator MD are mounted as close together as possible. Their critical components are the triode or tetrode transmitting valves and the thyratrons or other devices which generate the high power DC pulse. All of these had to be developed specifically for use in radar systems after the first equipments had been built with active components designed originally for use in communications transmitters and high power industrial welding equipment.

The receiver is a superheterodyne system, basically similar to those used in communications and TV, where the output from the local oscillator LO is mixed with incoming signals to produce the chosen and fixed intermediate frequency signal (IF), which is then amplified by a chain of fixed tuned amplifiers which determine the gain and bandwidth of the receiver. In a typical radar system the IF is in the range 20 to 45 megaherz. The gain is of the order 10^6 times. To permit amplification of microsecond pulses without unnacceptable degradation of their rise and fall times, the bandwidth is several megaherz. The signal to noise ratio (see Appendix 4), which is the true measure of the receiver's ability to detect signals of very small amplitude (microvolts) is determined primarily by the characteristics of the signal amplifier SA which precedes the mixer MX. It can be degraded by electrical noise generated in the signal amplifier or the mixer, or injected into the mixer by the local oscillator. Thus the critical active components in the receiver are the signal amplifier, and to a much lesser degree the mixer and the local oscillator. Initially, these were the same as the valves used in contemporary communications equipment, but greater impetus was given to their development by their radar applications. The head amplifier HA is the first stage of the IF amplifier chain. It and the signal amplifier are mounted next to the mixer so that the couplings between them can be optimised for maximum signal to noise ratio. [This condition is not necessarily the same as that for maximum transfer of signal from stage to stage]. Amplification of the IF signal before it leaves the receiver assembly makes it less prone to external interference as it is fed via a coaxial cable to the main IF amplifier A2. The penultimate stage in A2 is a diode rectifier which produces the video signal. This is fed via a pulse amplifier and a coaxial cable to the display D. [At this point the peak amplitude of the video signal is measured in volts or fractions of a volt.]

The critical component in the display is the cathode ray tube, and particularly the electroluminescent material forming its screen, which is required to give long afterglow in PPI applications. The beam deflection circuits in the display are triggered by a pulse from the modulator so that the start of the traverse of the spot across the screen is synchronised with the leading edge of the transmitted pulse. [In some systems, the trigger signal is generated in the display and controls the modulator].

2.2.2 Microwave Systems (above 1 GHz)

Twin aerial systems were also used in a number of Naval and Army microwave radar systems which were operational throughout the second world war. The receiving aerial was connected directly to the crystal mixer by a waveguide, and the transmitting aerial likewise to the magnetron transmitter. One advantage of the twin array system over the 'common T and R' arrangement shown at (b) was that it reduced significantly the problem of protecting the crystal from damage by the transmitted pulse and with life of the TR and TB cells, particularly when transmitter powers exceeded 50 kW peak. [In shipborne installations, there was always a finite probability that receiver performance might be impaired by high energy inputs arising from reflections from the superstructure or from the output pulse of a nearby transmitter].

In a 'common T and R' system, the paraboloid aerial P, which may be remote from the rest of the equipment, is connected to the magnetron transmitter MO by a waveguide system incorporating the TR and TB cells. TB is the transmitter blocking switch (at one time described as an Anti-TR cell or ATR), a gas filled device which is ionised by the RF pulse and so permits it to pass by with little loss of power. In its un-ionised state, it blocks the passage of signals reflected from targets, which would otherwise enter the magnetron and so reduce the effective sensitivity of the receiver*. The TR cell is also a gas filled device which is heavily ionised by the pulse as it passes. At the same time, it protects the mixer MX from damage and possible destruction by the transmitted pulse. In some systems, particularly later ones, the modulator provides a pre-pulse which is applied to the keep-alive electrode in the TR cell so that the cell is ionised before arrival of the RF pulse. This minimises the probability of damage to the mixer by RF energy passing through the TR cell during the early part of the pulse, when it might otherwise not be ionised sufficiently to give adequate protection. In its un-ionised state, it permits signals reflected from targets to enter the receiver with little attenuation. There is no amplification of the incoming signal, and so the signal to noise ratio of the receiver is determined primarily by the mixer. Because frequency conversion results in effective loss of incoming signal, the input from the local oscillator must have a low noise content. For the same reason the IF amplifier must have a low noise factor. Thus the mixer MX and the reflex klystron local oscillator RK, and to a lesser degree the head amplifier HA are critical components in microwave receivers. The reflex klystron and the crystal diode mixer were developed for radar applications and later used in communications equipment. The high power multicavity magnetron and the TR and TB cells were developed specifically for radar systems, as were the high power high voltage mercury thyratrons and triggered spark gap switches used in the modulator unit of landbased and shipborne installations, and the trigatrons and hard valve switches in mobile and aircraft equipment.

*In airborne systems such as AI MkVII, where the volume of the equipment had to be kept to the absolute minimum, the TB cell was omitted. Loss of received signals was minimised by arranging the RF circuit so that the impedance of the magnetron output circuit as seen from the TR junction simulated that of the missing TB cell.

Chapter 3

Metric Devices

3.1 Foreword

Metric radar systems were used in large numbers throughout the war. No microwave equipment was available at the time of the Battle of Britain. 200 MHz AI and ASV systems preceded and were then operated alongside their microwave alternatives. With the exception of systems designed to detect small targets at sea level, and some fire control equipments, Naval radar operated at or below 600 MHz, as did land-based early warning systems.

When the first RDF systems were conceived and developed at Orfordness and Bawdsey in the mid 1930s, they had for security reasons to be built with active components used by the television and communications industry in the manufacture of transmitters, receivers and displays, and by Signal School for use in in Naval communications. [No valve manufacturers had been 'allowed in' to RDF]. Receivers were modified television units. In transmitters, it was found that valves designed for CW operation could withstand pulsed anode voltages several times their DC ratings, and that their tungsten filament cathodes could be over-run to increase the available emission. Thus they could generate peak powers about ten times their CW output. This was adequate for experimental work, but it soon became obvious that improvements in performance would be obtained only by the development of active components designed specifically for use in radar systems, such as transmitting valves giving high peak power output in short pulses, and signal amplifier valves with high gain and low noise factors (see Appendix 4) for use in receivers with bandwidths wide enough to preserve the waveform of pulses with durations measured in microseconds.

Development of active devices was concentrated in three frequency bands – 50 to 100 MHz, around 200 MHz, and 600 MHz. A few transmitting valves were developed for use in experimental equipment operating around 1 GHz, but this activity was overtaken by the rapid advance of microwave technology, which could provide higher accuracy of bearing measurement with smaller aerial systems, and equal efficiency of conversion of DC power input into RF power output.

3.2 A summary of achievement

Apart from the initial availability of transmitting valves giving adequate power for the detection of ships and aircraft at ranges of a few tens of miles, operation at frequencies below 100 MHz provided early warning radar systems with immunity from the effects of cloud and rain. The Chain Home defence network, the MB2 mobile ground radar systems used by the RAF, and the Type 79 series of Naval equipments operated throughout the war at frequencies below 55 MHz. Type 281, the most powerful shipborne radar, operated at 90 MHz.

Below 100 MHz, receivers were designed to use 'general purpose' valves, (see Chapter 7). The early receivers were limited in performance above 50 MHz, but this deficiency was remedied in the late 1930s by the introduction of the 'Loctal' series by Philips Mullard, which extended the upper limit of high gain low noise amplifiers to above 200 MHz.

In transmitting valves, the requirements for high peak power output in pulses with durations up to 20 microseconds and repetition rates in the range 50 to 1,000 p.p.s caused significant changes in design. Anode dissipation (with a few exceptions) was no longer a limiting factor. Operation with high anode voltages, and the availability of high peak current, preferably from cathodes requiring low heating power, were prime desiderata.

The NT57 triode was the first valve developed specifically for use in radar transmitters. Designed in 1936 by H.G.Hughes of Signal School, it differed from their established silica envelope communications types in having a shorter anode (to reduce the output capacity and so increase the maximum frequency of operation), and twin 'hairpin' filaments to increase the available emission from the cathode. It was used in the Type 79 shipborne early warning radar, and in the early models of the MB2 mobile ground radar developed at Bawdsey.

The thermal efficiency of the cathode then became a limiting factor in design. [Under pulsed conditions, the heating power with pure tungsten filaments was 70 per cent or more of the total input to the valve]. Thus the introduction of thoriated tungsten filaments into NT57 by GEC and Signal School in 1939 marked a major advance in radar transmitter technology. [Thoriated tungsten filaments require about one sixth of the heating power of their pure tungsten alternatives. It had previously been accepted that only pure tungsten cathodes could be operated with the anode voltages required to give high peak power output.] A notable outcome of this innovation was the design of NT86 by F.M. Foley of Signal School in 1939. NT86 was the most powerful silica envelope valve built for use in radar systems. A pair operating as a push-pull oscillator was used in the Type 281 shipborne early warning radar to generate peak powers up to one megawatt at 90 MHz.

In addition to increasing the overall efficiency of transmitting valves operating below 100 MHz, the introduction of thoriated tungsten filaments made possible the development of thermally efficient smaller structures operating at 200 MHz for use in AI and ASV and in the CHL coastal defence network.

Because of their geometry and technology of fabrication, silica envelope triodes could not be scaled down in size to operate above 100 MHz. Each valve was individually fabricated and assembled, so that they were not suitable for production in quantity. Early in 1939, it could be foreseen that the demand for NT57Ts, which were used in the 50 MHz MB2 mobile ground radar systems as well as in Type 79 and its successors, would be beyond the combined capability of the manufacturing groups at GEC, Mullard and Signal School.

Both of these problems were solved by R. le Rossignol, W.H. Aldous and J. Bell of the GEC Transmitting Valve Group, who in 1938 developed a replacement for the NT57 based upon the MOV ACT10 television transmitter valve. VT58 had a pure tungsten filament, an external copper anode, and a side connection to the grid. It could operate efficiently at 200 MHz, and was therefore adopted for use in the development models of the CHL transmitters. Within a year, it was fitted with a thoriated tungsten filament. Type Approved as VT98, it was used in the production models of CHL and replaced NT57T in the MB2 mobile ground radar transmitter. By the end of 1941, several thousand had been manufactured by MOV. Le Rossignol's group also developed VT114, a metal-glass replacement for the NT77 tetrode developed by Signal School for the MB2 transmitter.

In 200 MHz receivers, development of the 'Loctal' RL7 signal amplifier by Mullard in 1940 provided a replacement for the 'acorn' tubes which had previously been the only types available for use in their input stages. It was much easier to manufacture in quantity, and superior in performance up to 300 MHz.

Extension of the frequency range to 600 MHz required major changes in technique.

At frequencies above 300 MHz, the inductance of the cathode lead in signal amplifier pentodes causes unacceptable degradation of their noise factor because of feedback of partition noise into the grid circuit. It also reduces the input impedance of the valve, which may result in reduction of the signal voltage on the grid because of the difficulty in designing an efficient coupling between the valve and its associated external circuits.

Thus development of the S25A series of 'grounded-grid' triode amplifiers by C.N. Smyth and J. Foster of the Receiver Group at STC Ilminster in early 1941 was a major contribution to receiver technology. [In the grounded-grid mode, the input signal is applied to the cathode. The grid acts as a screen between the input and output circuits. There being no second grid, there is no degradation of noise figure by partition noise]. It was found that S25A, later known as CV16, was superior in performance to the RL7 pentode at 200 MHz. This prompted Mullard to develop RL37, a grounded-grid triode on a Loctal base, which was marginally inferior in performance to S25A but more suited to the geometry of 200 MHz circuits.

In transmitters, the shift to 600 MHz required major changes in fabrication technology to minimise the adverse effects of lead inductance and to reduce inter-electrode spacings. Thus development at GEC in 1939 of the first of the 'Micropup' series of transmitting valves by W.H. Aldous and J. Bell, and of the family of grounded-grid triodes by G.W. Warren were major advances in metric transmitting valve technology.

All of their designs used disc or ring seals for the anode and grid connections. Their operation depended equally upon the intoduction of oxide cathodes. [As in the case of thoriated tungsten filaments, it had been assumed that oxide cathodes would not withstand high anode voltages because of the risk of destruction of their surface by ion bombardment or flash-over. The GEC group found that this could be overcome by 'anode modulation' (see 6.3 and 6.4) where the input voltage is applied only for the duration of the pulse.

NT99, the most widely used of the micropups, was manufactured in quantity for use in the AMES11 50 cm mobile ground radar transmitter, in Naval fire control systems, and in airborne 200 MHz equipment. The CV90 triode and its amplifier variant CV153 were alternatives to the STC series, which also included an oscillator.

The development of metric duplexers required no major advances in technology. Perhaps the most interesting innovation was made by H.J. Reeves of Signal School, who used paralleled television diodes (D1s or EA50s) as the switches in the duplexing system of Type 79B, the first Naval application of 'common T and R' working, and subsequently in Type 281.

3.3 Transmitting Valves

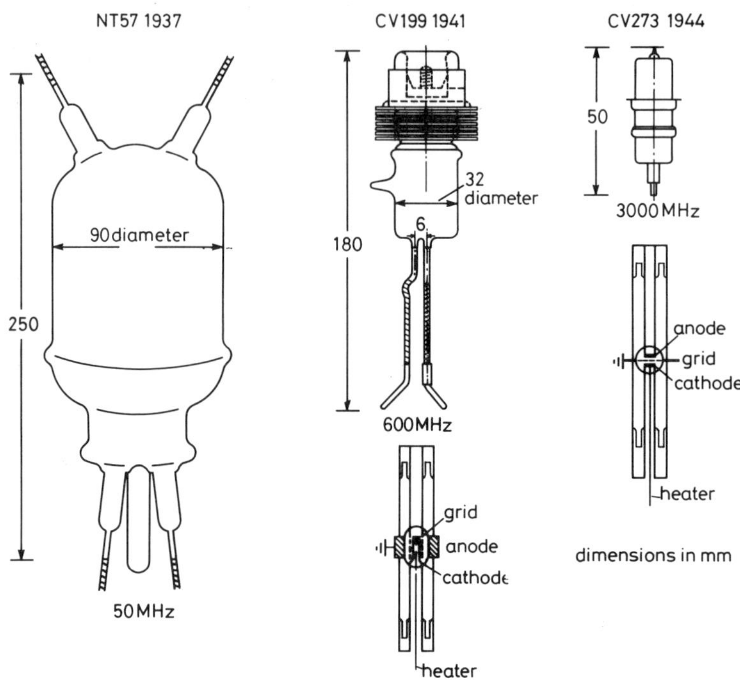

3.3.1 Foreword

Transmitting Valves are designed and operated to give high efficiency of conversion of the DC input power to RF output. To obtain maximum efficiency, they are biassed well beyond cut-off of the anode current by a negative voltage on the grid, and conduct during an 'angle of flow' equal to about one third of a cycle of the RF input. Thus the greater part of the anode current flows when the anode voltage is close to its minimum value, so making the conversion efficiency high (theoretically around 75 per cent). The anode load is a resonant circuit, often described as a 'tank' circuit, which acts as an energy store from which the RF output is fed to an aerial system or the next amplifier stage. Intrinsic losses in the tank circuit may reduce the overall efficiency by up to 10 per cent.

Because the grid is driven positive during the conducting period, it will collect a fraction (typically about one seventh) of the current emitted from the cathode, so reducing the available output and requiring that the grid be prevented from overheating. The power required to drive the grid positive is provided by the preceding stage, being typically less than 6 per cent of the amplifier output. Being automatically biassed and driven by the RF input to their grids, which is part of the available output power, self-oscillators have efficiencies lower than those of amplifiers by between 5 and 10 per cent. Thus the overall efficiency of a self-oscillator driving an amplifier with a power gain typically greater than 10 times is not significantly different from that of an oscillator providing the same output to an aerial system. The combination has the advantage that its frequency of operation will not be affected by the variation of input impedance with frequency which is characteristic of some multiple element aerial systems. Its disadvantages are duplication of tuned circuits and possibly power supplies, the greater number of components, and its larger volume.

It can be shown that for an angle of flow around 140°, and taking account of the fraction of the cathode current collected by the grid, the available RF output power is equal to $\frac{1}{2}I_f(V_{DC}-V_{min})$, where V_{DC} is the DC anode voltage, and V_{min} the minimum value of the anode voltage (typically a small fraction of V_{DC}). I_f, the fundamental component of the anode current waveform, is approximately one third of I_m, the maximum permissible cathode current. Under these conditions, I_m is roughly five times the DC component of the anode current, which determines the DC input power. Thus transmitting valves are designed with cathodes having a reserve of emission large enough to produce at least five times the DC anode current without becoming temperature limited. This ensures that there is no limitation of power output by 'clipping' of the peaks of anode current, and that the emission is always space charge limited, i.e. that the cathode is surrounded by a sheath of electrons which minimises the probability that its surface will be damaged by positive ion bombardment. [Because it is immune to bombardment, a pure tungsten cathode can be operated so that it is close to saturation at the peaks of the anode current. This advantage is offset by its high operating temperature (see 3.3.5 and Appendix 6).

To obtain maximum output (subject to the limitations of anode dissipation and available peak cathode current), transmitting valves are designed to operate with the highest DC anode voltage which their geometry will permit. [There must be no loss of control of the anode current; the permissible maximum bias voltage is determined by the grid-cathode geometry.]

Tetrodes can operate with higher anode voltages than triodes with the same grid-cathode spacing because the screen grid (G_2) limits the penetration of the anode field into the grid-cathode space. In the driven condition, G_2 assists the flow of electrons from the cathode to the anode, so that the grid driving power is less for the tetrode. The anode current is reduced by collection of electrons by the screen grid , but this can be minimised by alignment of the grids, as in radial beam tetrodes which, being cylindrical, require no 'beam forming' plates. In addition to providing higher gain and output power, tetrode amplifiers have the advantage over triodes in requiring no neutralisation of positive feedback from the output to the input circuit through the anode-grid capacitance*, which could otherwise cause uncontrolled self-oscillation (G_2 is at RF earth potential).

In triode amplifiers, the required negative feedback can be obtained by using a pair of valves in push-pull and cross-connecting their anodes and grids by a pair of adjustable capacitances with a maximum value slightly higher than the inter-electrode capacitance. (In push-pull, the RF inputs to the grids are in anti-phase). Alternatively, triodes can be operated in the 'grounded-grid' configuration, where the input is applied to the cathode and no neutralisation is required.

As self-oscillators, where power gain is less important, tetrodes lose some of their operational advantage over triodes. They also have the disadvantages that they cost more to produce and require an extra high voltage supply. Because of the problems involved in constructing disc seal assemblies with more than one grid, their use was limited to frequencies below 100 MHz. Triode self-oscillators were used exclusively as the transmitters in metric radar systems operating at and above 200 MHz.

The transmitting valves in radar systems are turned on for periods between one and twenty microseconds at intervals between one and twenty milliseconds. Thus the average power input to the anode is of the order of thousandths of the peak value. Under these conditions, anode dissipation is not a limiting factor, so that the grid and anode voltages can be raised to the limits imposed by immunity from flash-over or loss of control by the grid bias voltage. In early radar systems, the operating anode voltage of the transmitting valves under pulsed conditions could be raised to between five and seven times their CW rating. This, combined with over-running their tungsten filaments*, resulted in generation of peak powers ten times higher than their CW output.

*For a pure tungsten filament, an increase in operating temperature from 2,200°K to 2,500°K increases the available emission by about 20 times. The corresponding increase in cathode heating power is about 1.7 times. The higher operating temperature causes a significant but not unacceptable reduction in life.

Under pulsed conditions, the cathode heating power of transmitting valves with pure tungsten filament cathodes is a large fraction of the total power input, (typically between 0.7 and 0.9). Thus there is much to be gained by reducing the heating power to a minimum. This will simplify thermal design of the valve itself and its cooling system, and reduce significantly the power required to operate the transmitter, which is particularly important in mobile and aircraft equipment.

The introduction of thoriated tungsten filaments by GEC and Signal School in 1939 was a major contribution to metric radar technology (see 3.3.5). It had previously been assumed that only pure tungsten cathodes could withstand high anode voltages, even though thoriated filaments had been used for several years in communications transmitting valves operating with up to 5 kV HT supplies. [A thoriated filament giving the same emission as a pure tungsten alternative will have about one third of the area and require less than one sixth of its heating power].

Other important innovations made by the GEC transmitting valve group were in the use of oxide cathodes with higher anode voltages than previously thought possible, and in the adaptation and development of their metal-glass envelope technology, which together made possible the design and production of pulsed transmitting valves generating tens of kilowatts peak at frequencies up to 1 GHz (see 3.3.6)

3.3.2 Early Days

By 1937, transmitting valves in television and civil and military communications systems were generating hundreds of kilowatts output at frequencies up to 50 MHz. Smaller water-cooled triodes could produce a few tens of kilowatts at frequencies up to 100 MHz. The silica envelope valves developed and made by Signal School for shipborne Naval communications transmitters generated powers measured in kilowatts at frequencies up to 50 MHz, and were much smaller and more robust than their glass envelope equivalents.

The first experimental early warning radar systems built at Bawdsey in 1937 used silica envelope triodes (NT46 and NT57) with their filaments over-volted to increase the available anode current. The transmitters built by MetVick for the Chain Home defence network used water cooled continuously evacuated demountable tetrodes which they had developed and manufactured for high power broadcast and communications transmitters. They were fitted with tungsten filament cathodes (the only type that could be used in demountable systems) and generated 40 kW mean power output at 50 MHz. In CH, they gave peak powers around 600 kW in pulses with up to 35 microseconds duration at intervals of 20 milliseconds (50 p.p.s). The filament temperature in the output stage was higher than that for CW operation, being set to give lives between 1,000 and 1,200 hours before replacement.

3.3.3 Silica Envelope Valves

The silica envelope valves made by Signal School were the obvious choice for use in shipborne and land based mobile radar transmitters, partly because of their small size in relation to their output, and partly because development to increase peak power output would be automatically protected by military security. [Before 1938, no valve manufacturers had been cleared for access to information on RDF]. NT46s or NT57s driving a pair of neutralised NT57s were used in the early models of the MB1 and MB2 mobile ground radar systems designed at Bawdsey in 1937 for operation by the RAF, and in the GL1 and GL2 gun-laying equipments developed soon after. A pair of NT57s operating as a push-pull self-oscillator was used in Type 79, the first shipborne early warning radar. Built in 1938, it transmitted up to 15 kW peak at 40 MHz in pulses of 5 to 20 microseconds duration with a recurrence rate of 50 p.p.s.

Designed in 1936 by H.G. Hughes, leader of the Signal School valve group, NT57 was the first valve intended specifically for use in radar systems, (NT46 was an established Naval communications type). NT57 differed from NT46 in having an anode half its length, which reduced the output capacitance and so increased the maximum frequency of operation. The single 'hairpin' filament cathode of NT46 was replaced by a pair of shorter and thicker filaments, which were connected in series so that they could operate with the same voltage. Rated at 15 V 48 A, the cathode in NT57 could produce about 5 A saturated emission, over three times that of the NT46. It was overvolted by 10 per cent in Type 79.

Early in 1937, following a suggestion from the ground radar group at Bawdsey, the Signal School valve group designed NT60, a tetrode with a tungsten filament. The grids were not aligned. Samples were available in December 1937. Driven by a pair of NT57s, they were used in the early production models of the MB2 developed by MetVick in 1939 and manufactured from early 1940. Rated at

18 V 130A, NT60 had a much larger cathode than NT57. The peak power output from a pair was of the order 100 kW, at least four times greater than that of the triode with the same DC anode voltage.

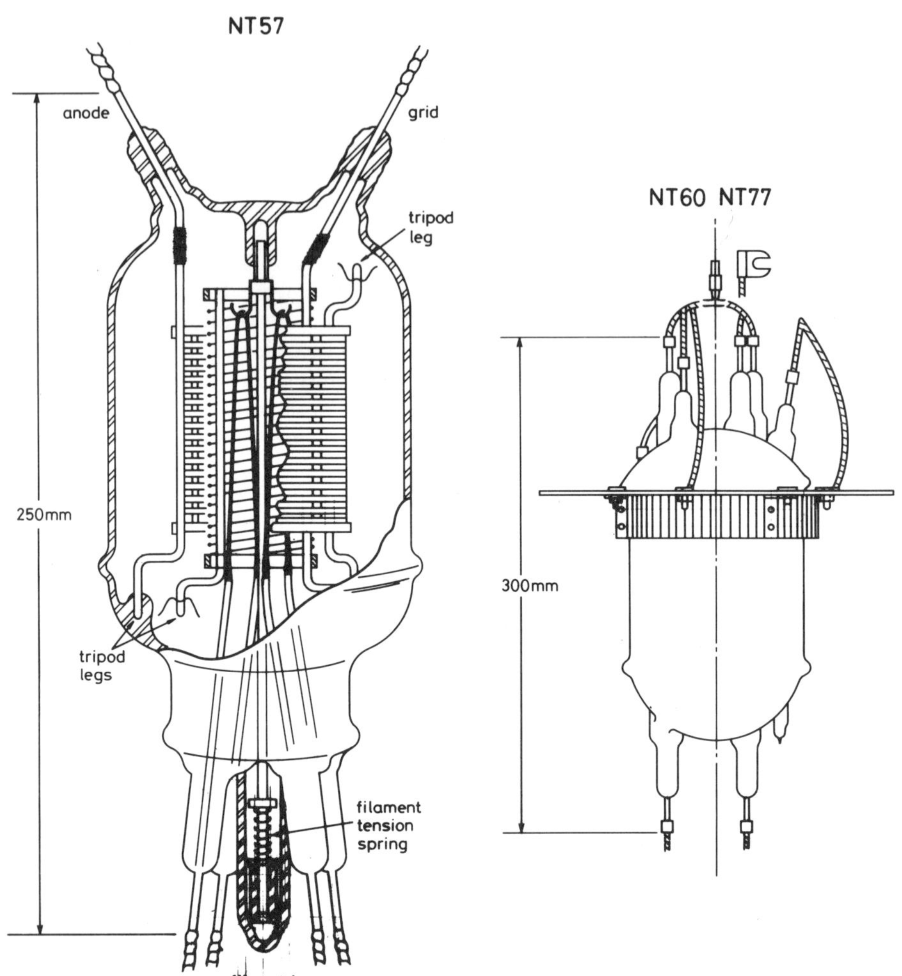

Silica envelope valves used molybdenum electrodes and tungsten rods with flexible braid extensions as the connections to the electrode system. Because of the disparity between the coefficients of thermal expansion of tungsten and silica, it was impossible to make vacuum tight seals between them. For this reason, the valves made up to 1939 used a technique developed by B.S. Gossling, S.R. Mullard and T.E. Goldup at Signal School in the early 1920s. Each rod emerged from the envelope through a loose fitting silica tube which was slightly flared at its outer end and filled with lead, which wetted both materials to make a vacuum tight seal and could accommodate the difference in thermal expansion. [The seals required forced air cooling]. The overall length of the connecting leads limited operation to frequencies below about 50 MHz. In 1939, Philips in Eindhoven and GEC developed independently a 'graded' seal, in which short lengths of glass tube with different coefficients of expansion were sealed together to form a sub-assembly which effectively matched tungsten at one end and silica at the other, so that vacuum tight seals could be made without using lead. The consequent reduction in the length of the leads raised the limit of operation to about 100 MHz.

In 1939, F.M. Foley of Signal School designed NT86, a development of NT57T with a longer anode (85 mm cf 70 mm) and three thoriated tungsten hairpin filaments instead of two. With a total rating of 10.5 V 100 A, the available peak emission was 90 A. To take full advantage of the increased filament area, the tops of the hairpins were rounded and supported by yokes attached to the filament tensioning rod. This made each leg of the hairpin practically vertical, so that the effect of the grid bias and drive voltages was much more uniform along the length of the anode. [In the original single and twin hairpin designs, the apex, being more remote from the grid than the lower parts of the legs, was more effectively screened from the anode field, and so contributed a much smaller fraction of its available emission to the total anode current]. To accommodate the larger filament assembly, the diameters of the grid (40 mm) and anode (70 mm) were larger than those of NT57. The grid itself was similar in that it was a spiral of 0.3 mm wire with the turns spaced 3 mm apart. Installed in the Type 281 early warning radar in 1940, a pair operating as a push-pull oscillator generated peak powers up to one megawatt peak at 90 MHz in pulses of 2 to 3 microseconds duration at a recurrence rate of 50 p.p.s. The peak power input was up to 60 A at 28 kV. When the transmitter was operated with 15 microsecond pulses, the peak power output was limited to 300 kW by the permissible maximum anode dissipation. NT86 was the last and probably the most important development of silica envelope transmitting valves.

3.3.4 Metal-Glass Envelope Valves

Early in 1939, it became increasingly obvious that production of silica envelope valves would be unable to keep pace with the predicted numbers required for mobile ground radar stations and CHL, even though the manufacturing groups at GEC, Mullard and Signal School were working to their full capacity.

Each valve had to be fabricated and assembled by hand. The anodes were cylindrical open ended baskets, made from molybdenum strip about 2 mm wide woven on a mandrel over three vertical molybdenum rods attached to rings at each end of the basket. The rods extended beyond the rings to form tripod supports at each end of the anode. One leg was joined to the anode connector, a tungsten rod sealed through the top of the bulb. The others were fitted into sockets moulded into the upper and lower cups which housed the connectors. These had flexible braid extensions brazed to their outer ends. The grid was constructed in a similar way, being wound as a spiral over the three support rods, with each turn laced to each rod by a fine tungsten wire which spiralled round it. No welding was used for joining parts within the envelope. The support rods were rivetted to the rings, which were machined from the solid. Junctions between the support rods and the connectors were made by lapped joints wrapped with fine tungsten wire. A similar technique was used for joining the filaments to their connectors, except that these were ground flat for drilling and the filaments hooked through the holes before wrapping. To ensure that there would be no electron bombardment of the envelope, the 'cold ends' of the filament were shielded by the grid, which was longer than the anode. [If electrons reach the glass, the effect of secondary emission is to charge it positive, which could cause internal flashover or puncturing of the bulb by external sparking to nearby metal parts.] The tops of the hairpin filament were supported by hooks attached to a rod passing through the axis of the assembly with its upper end located in a socket moulded into the upper cup and its lower end in a sealed-on sub-assembly which included a compression spring to force it upwards and so prevent undesired movement of the filaments. The seal assemblies were mechanically fragile. Production of the silica parts was difficult and expensive (there was only one supplier). Thus silica valves were not suitable for quantity production.

This problem was solved by R. le Rossignol and other members of the GEC Transmitting Valve group at Wembley (including B.S. Gossling, who had joined them several years previously), who adapted and developed technology used in their 'CAT' (copper anode triode) range of broadcast and communications valves and produced replacements for NT57 and NT60, which were manufactured in quantity by MOV.

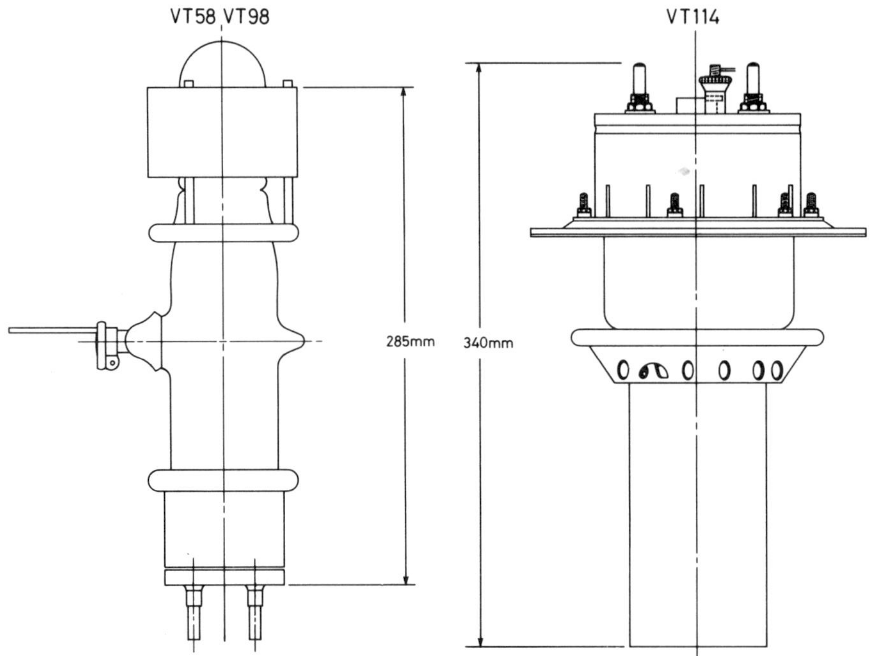

E960, later known as VT58, was first used in the experimental 200 MHz CHL transmitters made by the Army research group at Bawdsey in 1938. Derived from the ACT10 television transmitting valve, E960 was superior both mechanically and electrically to NT57 (its connecting leads were shorter and its seal assemblies much more robust). From early 1940, it replaced the silica valve in the driving stage of the MB2 ground radar transmitter and in the driving and output stages of GL1 and GL2. As amplifiers, a pair of VT58s gave 30 kW peak in 3 microsecond pulses with a recurrence of 1,000 p.p.s, about one and a half times the output of NT57s in the same equipment. By the end of 1941, several thousands had been manufactured.

At the first CVD meeting in January 1939, there was a lengthy discussion of the possibility of making a direct replacement for the NT60 tetrode, including the use of a glass envelope, which was quickly rejected as impractical. GEC then developed E1024, using fabrication techniques similar to those of E960. Within a few months, E1024 was modified to include a thoriated tungsten filament. The first samples were tested successfully in August 1939. In November, they were fitted experimentally as alternatives to the silica NT77 in MB2 transmitters, and replaced the silica valve in 1940. [Following the successful trials of thoriated filaments in E1024, they were fitted to NT60 which was then renumbered as NT77, later known as CV50].

No records of the internal construction of VT58 and VT114 have been located. The filament power rating of VT58 (12.6 V 58 A) and NT57 (15 V 48 A) were almost equal, but the higher power output of VT58 shows that more effective use was made of the available emission. The length of the anode of VT58 was significantly less than that of NT57; its diameter was about the same. Thus one may deduce with some confidence that the internal geometry of VT58 was different from that of NT57, particularly in the construction of the cathode. It is almost certain that this was a tungsten spiral similar in form to that used soon after in VT90, the first of the 'micropups'. There was little difference between the filament ratings, the output powers, and the anode dimensions of VT114 and NT60. This, and the short time taken to produce the prototype VT114s, indicates that their internal geometries were basically identical.]

Despite the success of the GEC developments, Naval radar transmitters continued to use silica envelope triodes. This was partly because self-oscillators could operate with the required efficiency and frequency stability when coupled to shipborne aerial systems so that no buffer stage was necessary, and partly because modifications to shipborne equipment to accommodate VT58 metal-glass triodes in place of NT57s would have been operationally inconvenient, if not worse. Apart from that, the Navy had confidence in the reliability of silica valves based on experience over many years of operation in communication systems. With demand limited to Naval requirements there was no great difficulty in maintaining supplies.

3.3.5 Thoriated Filaments

The first use of thoria in tungsten filaments was to retard crystal growth in lamp coils, which limited their operating lives. Years later it was found to have a beneficial effect on electron emission. A monolayer of thorium on tungsten reduces the work function* from 4.5 to 2.7 volts (see Appendix 6).

*The work function of a surface is a measure of the energy that must be given to an electron to enable it to escape from that surface, and is expressed in 'electron volts'.

A thoriated filament at 2,000°K will give a theoretical temperature limited saturated emission between 2 and 3 amperes per cm^2, roughly ten times that of a pure tungsten filament at 2,500°K. The corresponding heating powers are 30 W and 70 W per cm^2. In practice, the thermionic emission from thoriated tungsten is limited to about 1 A per cm^2, so that a typical thoriated tungsten filament will have less than one half of the area and require less than one fifth of the heating power of its pure tungsten alternative. Thus thoriated filaments can be used in much smaller valves, or be used to obtain higher peak emission in valves with otherwise similar geometry.

Valves with thoriated tungsten filaments were used in broadcast and communications transmitters from the mid 1930s, but it became accepted that they could not be operated with more than a few kilovolts applied to the anode, even though processes had been invented for producing a layer of tungsten carbide at the surface so that loss of thorium from the surface would be balanced by reduction of thoria. [This could be explained by the difficulty of obtaining and maintaining high vacua in transmitting valves (the surface layer of thorium is easily removed by ion bombardment)].

At the CVD meeting in January 1939, economy of cathode heating power was given very high priority. In late 1938, T.J. Jones of Signal School had experimented with thoriated tungsten filaments in NT57. In April 1939, the first samples of NT57T (NT57 with a thoriated filament) were made and tested by Signal School in the Type 79 early warning radar, and by the Army group at Bawdsey in an MB2 transmitter. Subsequently, it was found that lives up to 7,000 hours could be obtained with the valves operating at 15 kV. [In March 1939, GEC, who were making NT57s as a second source to Signal School, used eddy current heating (see Appendix 6) to outgas the electrode assemblies of NT57Ts while they were on the pump. Eddy current heating was a standard method used by manufacturers of receiving valves and transmitting valves with glass envelopes; Signal School had preferred to use electron bombardment from the cathode, which was acceptable with pure tungsten, but a more complicated and lengthy process with thoriated filaments]. By September 1939, NT57T was in small scale production at Signal School and MOV. Its heating power, about 300 watts, was two fifths that of NT57. Its power output was 50 per cent greater.

In June 1939, Signal School tested thoriated tungsten filaments in NT60, and supplied samples for test at Bawdsey in September. Shortly before, GEC had successfully tested the first samples of E1024 (VT114) at 28 kV. As a result of these tests it was agreed that NT60, [which was being made in small numbers at MOV and Signal School] would be fitted with thoriated tungsten filaments and renumbered as NT77. The filament ratings of VT114 and NT77 were approximately 700 watts, less than one third that of NT60. The power output from a pair in push-pull driven by NT57s was up to 400 kW peak at 50 MHz, about double that of a pair of NT60s. Its specified maximum frequency of operation was 60 MHz.

Following the success of VT114, GEC fitted thoriated filaments in E960 (VT58). The first samples were available for test in September 1939. In 1940, E960T, then known as VT98, replaced VT58 in MB2 and CHL transmitters. Its filament rating, 400 watts, was about two fifths, and its peak output, 100 kW, twice that of VT58.

3.3.6 Transmitting Valves at 600 MHz

In 1939, W.H. Aldous and J. Bell of the GEC Transmitting Valve Group designed E1046 (VT90), the first of a series of 'micropup' metal-glass envelope valves intended for use at frequencies above 200 MHz. (The frequency limit of VT98 was about 300 MHz).

All of the valves in the series used the same form of construction. The anode was a short length of copper tube, flared and tapered at each end as a preparation for attachment of the glass parts of the envelope by Housekeeper seals (see Appendix 6). The grid was a 'parrot cage' comprising a number of short molybdenum or tantalum wires spaced uniformly round and welded to a shallow dish at one end and a ring at the other. The dish was in good thermal contact with a flared and tapered copper cup, to which it was attached by a central screw after a glass tube had been sealed to the flare. The first stage in the assembly process was to seal the grid glass to the anode glass while the anode and grid were held in a jig which ensured their accurate concentricity. The cathode in VT90 was a thoriated tungsten spiral filament supported by tungsten rods sealed through the base of a glass cylinder, which was attached to the anode flare to complete the assembly.

NT90, which generated peak powers up to 5 kW, was used only in the experimental models of 200 MHz AI and ASV equipment. Requirements for higher power and operation at 600 MHz were met by radial enlargement of the electrode system and replacement of the thoriated tungsten spiral filament by a cylindrical oxide cathode*. [This required that the valve be anode modulated (see 6.3 and 6.4) because of the high probablility that flash-arcs would damage the cathode if DC voltages equal to the voltage during the pulse were applied for more than a few microseconds.]

*An oxide cathode (see Appendix 6) has a work function less than 2 volts, so that its saturated emission at 1,000°K is theoretically 14 A per cm². Its radiation loss is approximately 2 W per cm², so that an oxide cathode is theoretically at least ten times more efficient than a thoriated tungsten filament. In practice, the DC emission from an oxide cathode is limited to about 1 A per cm², but it was found that the emission density with pulses measured in microseconds could be several times higher. Because they operate at much lower temperature, and have continuous surfaces, they can be used with close grid-cathode spacings, and so make possible the design of valves to operate with high thermal and electrical efficiency at frequencies above 200 MHz. For geometric reasons, a cylindrical cathode with internal heater has higher thermal efficiency than an assembly of filaments giving the same emission.

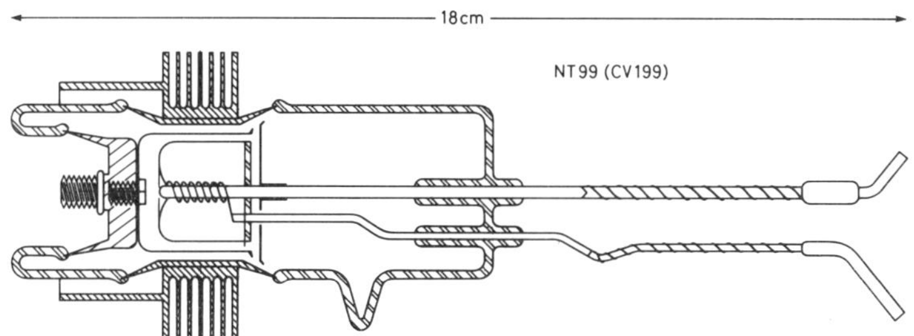

E1232 (NT99), developed early in 1941, was the most widely used Micropup. Its cathode was a nickel cylinder, closed at both ends and heated by a central tungsten spiral. Samples were available for test in April. In July 1941, it was being made in small quantities (about 8 per week). Tests at 200 MHz had shown that a push-pull oscillator pair would generate 150 kW peak. By January 1942, it was in production at MOV for use in AMES11 600 MHz mobile radar systems operated by the RAF, and in 200 MHz AI and ASV equipment. The power output from a push-pull oscillator was 100 kW peak with an efficiency around 45%. In Naval gunnery radar systems they generated powers around 150 kW. (Powers up to 200 kW had been obtained experimentally). Its heating power (6 V 6.5 A) was roughly one seventh that of VT98, which gave the same output at 200 MHz. [This illustrates the advantage gained from the use of a cylindrical oxide cathode].

Micropups were designed as oscillators, and were operated in the 'grounded-anode common-anode' mode, where the anode is at RF earth potential and connected to the outer conductors of a pair of coaxial lines; the grid is connected to one inner conductor, the cathode to the other. In the push-pull circuits used in AMES11, the anodes were strapped together and twin transmission lines connected between the two grids and between the two cathodes.

In 1943, following the success of their CV90 grounded grid oscillator (see 3.4.5) the GEC team designed CV288, a power amplifier intended for CW and pulsed operation at and above 600 MHz. Tested at TRE, it generated 50 kW peak output with a power gain of 4 times and an anode efficiency of 50 per cent. It could also operate as a frequency doubler, giving 16 kW output with a gain of 2 times. Its anode voltage was limited to 5.5 kV by external breakdown. Although a few experimental crystal controlled transmitters using CV288 as a multiplier and as the output amplifier were built at TRE in 1945, they were not used operationally.

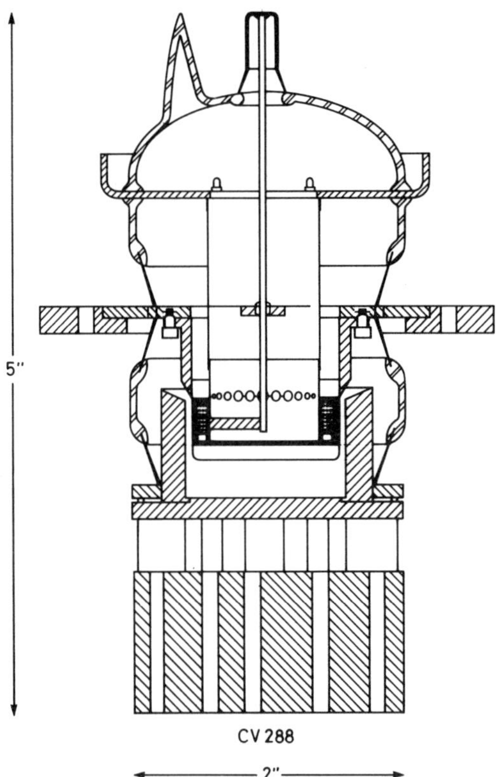

3.4 Receiving Valves

3.4.1 Foreword
By 1937, the requirements of television and civil and military communications

had resulted in the development of superheterodyne receivers with bandwidths of a few megaherz and sensitivities measured in microvolts at frequencies up to 50 MHz. In their 'front ends', British receivers used general purpose valves such as the Mazda SP41 as signal amplifiers up to about 30 MHz. Direct input to

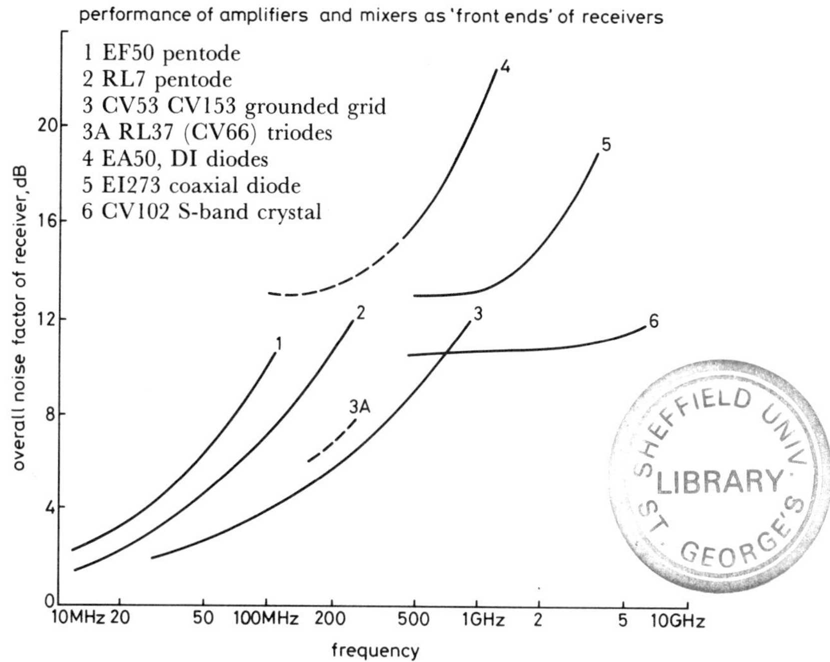

'television diode' mixers, such as the Mullard EA50 and Mazda D1, was used at frequencies up to 100 MHz (with some loss in sensitivity). In 1938, the introduction of the EF50 pentode by Mullard provided RF amplification up to 100 MHz. In 1940, they produced RL7, a variant with higher transconductance and aligned grids to minimise partition noise (see Appendix 4), which extended the range of low noise RF signal amplifiers to 300 MHz. Thus receivers in the early radar systems, such as CH, GL1, and Type 79 could be built with commercially available 'general purpose' receiving valves. Type 281, which operated at 90 MHz, used RL7. At 200 MHz, the first AI, ASV and CHL receivers used imported American 'acorn' tubes.

3.4.2 Signal Amplifiers at 200 MHz

At frequencies above 100 MHz, the length of the leads between the electrodes and their associated circuits causes significant and continuous degradation of the noise factor of pentode signal amplifiers. The impedance of the cathode lead, being common to the input and output circuits, provides the means whereby feedback causes a reduction in the input impedance of the valve, while shot noise in the anode current and partition noise in the screen current are injected directly into the grid circuit.

RCA in America were the first to extend the frequency range of high sensitivity receivers to around 500 MHz. Their 'acorn' tubes, developed in the mid to late 1930s, used short radial connections to active modules much smaller than their Octal equivalents, and bulbs which by their size and configuration gave their name to the series. The bulbs were made in two parts, each with a flange, and sealed together to sandwich the radial leads.

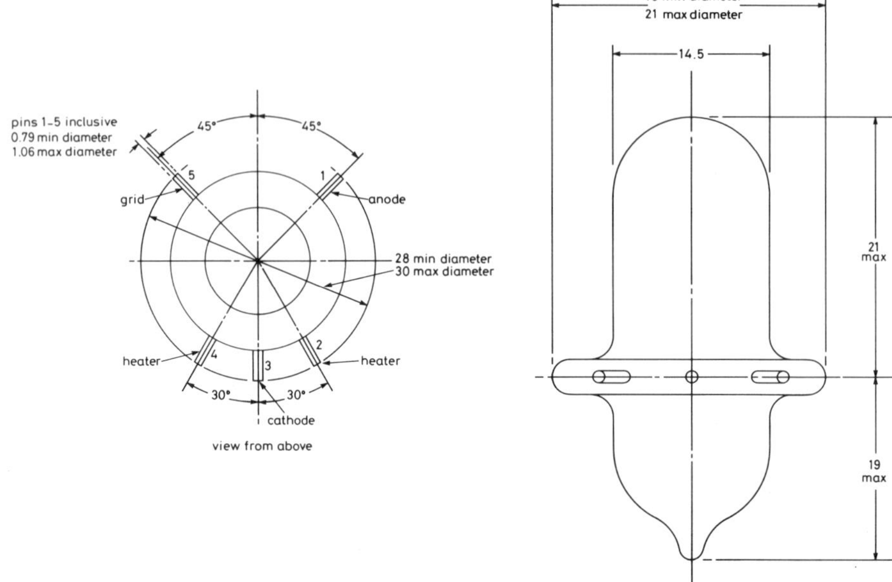

dimensions in mm

Early British 200 MHz radar receivers used imported acorn tubes, which were copied with difficulty by MOV and Mullard. They required highly skilled personnel to fabricate the components and to assemble each unit individually before the sealing operation, and it was soon found that the standards required could be maintained for at most a few months. Training took several weeks, so that production had to be backed by a continuous and labour intensive training programme. Development of the RL7 by Mullard in 1940 provided a solution to the problems of maintaining supplies.

RL7 was a variant of EF50, having duplicated cathode leads to reduce the common impedance effect, and aligned grids to reduce partition noise. With a noise factor around 11 dB at 200 MHz, its performance as a low noise signal amplifier was superior to that of acorn pentodes at frequencies up to 300 MHz. At about the same time, GEC developed a mechanically different tube with performance marginally superior to that of RL7, but after tests on a number of tubes had shown that the performance of the two types was not significantly different, GEC agreed that the RL7 be preferred to avoid duplication. They would 'second source' Mullard as suppliers of EF50 (CV1091) and RL7

(CV1136), also known as VR91 and VR136. [Acorns were black listed by ISTVC in June 1941].

After the demonstration by Smyth's group at STC that the CV53 disc-seal grounded-grid triode (see below) could give an improvement of about 4 dB in the noise factor of 200 MHz receivers using pentode signal amplifiers, Mullard developed RL37, a grounded-grid triode with similar characteristics on a Loctal base. Placed in front of an RL7 pentode it improved the overall noise factor by about 3.5 dB. Similar results were obtained with the E1305 triode developed by GEC at about the same time. As before, GEC agreed that RL37 be the preferred type, which was Type Approved as CV66 in 1942.

3.4.3 Signal Amplifiers at 600 MHz

At frequencies above 200 MHz, the adverse effects of the cathode lead inductance on the input impedance and noise factor of pentode signal amplifiers increase rapidly, so limiting their use to frequencies below 400 MHz. [Reduction of the input impedance may require that the coupling between the valve and the input circuit of the receiver be designed as a step-down transformer, so reducing the amplifier input voltage and degrading the noise factor].

A solution to this problem was obtained through the development of 'grounded-grid' triode amplifiers by the team led by J. Foster which was part of C.N. Smyth's receiver group at STC Ilminster. Grounded-grid power amplifiers had been used previously in broadcast and communications transmitters to eliminate the need for 'neutralising' positive feedback of signal from the output to the input circuit through the anode-grid capacitance of the amplifier valves, which could otherwise cause uncontrolled self oscillation in grounded cathode systems. It is believed that the suggestion to use the same technique in signal amplifiers came from C.E. Strong, Chief Engineer of the STC Radio Division.

In the grounded-grid configuration, the input signal is applied to the cathode and the grid acts as a screen between the input and output circuits. There is some penetration of the anode field into the grid-cathode space, but this has a negligible effect if the triode has a high 'amplification factor'*. There being no screen grid, there is no degradation of noise factor by partition noise. It can be shown that provided the amplification factor is high (around 100), the input impedance of a grounded-grid triode is the inverse of its transconductance*. Because the same signal current flows in the cathode and anode circuits, the voltage gain is the ratio of their impedances. If the gain is greater than times ten, the noise contribution from a second stage can be neglected, so that a typical front end amplifier with a transconductance of 5 mA/V and an anode load impedance over 2000 ohms will determine the overall noise factor of a receiver.

---

*The amplification factor of a triode is the ratio of the anode voltage and corresponding minimum grid voltage at which no anode current flows, and is thus a measure of the penetration of the anode field into the grid-cathode space. Its transconductance is the measure of its ability to amplify input signals by variation of the anode current, being measured at a fixed anode voltage as the change in anode current per unit change of grid voltage, and expressed in milliamperes per volt. It can be shown that the dynamic anode resistance of a triode (as opposed to its static resistance, which is the ratio of the anode voltage to the operating anode current) is the ratio of its amplification factor to its transconductance. Thus its theoretical maximum voltage gain is equal to its amplification factor. In practice, the gain will be reduced by the impedance of the external load, which shunts the anode resistance of the amplifier itself.

Thus grounded-grid signal amplifier triodes must be designed with the minimum anode-grid capacitance (to maximise the impedance of the output circuit), to operate with the minimum current necessary to give a high transconductance (to reduce shot noise), and have the maximum compatible amplification factor. [The two characteristics are inter-related geometrically]. The grid must be as close as practicable to the cathode to maximise transconductancee, (It is well known that the optimum spacing of the grid wires is approximately equal to the grid-cathode separation), but not so close that heating of its surface by thermal radiation from the cathode could cause emission of electrons from particles evaporated on to it from the cathode during processing and operation of the valve. Use of the minimum grid-cathode separation reduces the increasingly adverse effect of electron transit time on performance at frequencies above 1 GHz.

In 1939, Smyth's group was given the task of developing a 600 MHz receiver which would match the performance of available 200 MHz units with acorn RF pentodes in the input stages. After the group had tested samples of a beam deflection tube supplied by GEC, and finding that they were unacceptably noisy, [there is no space charge smoothing of shot noise in the beam], Foster and his team developed a family of grounded-grid triodes.

[During tests on the 200 MHz receivers, Smyth differentiated between noise generated in the input circuit and that contributed by the input amplifier. His method was to place the receivers in a refrigerator and measure their noise factors as a function of temperature. [Noise in passive components will be zero at 0°K. Shot noise carried by the anode current of an amplifier is independent of ambient temperature]. Later on, Smyth identified 'total emission damping', a transit time effect which causes a marked and rapidly progressive reduction in the input impedance of amplifiers operating above 2 GHz].

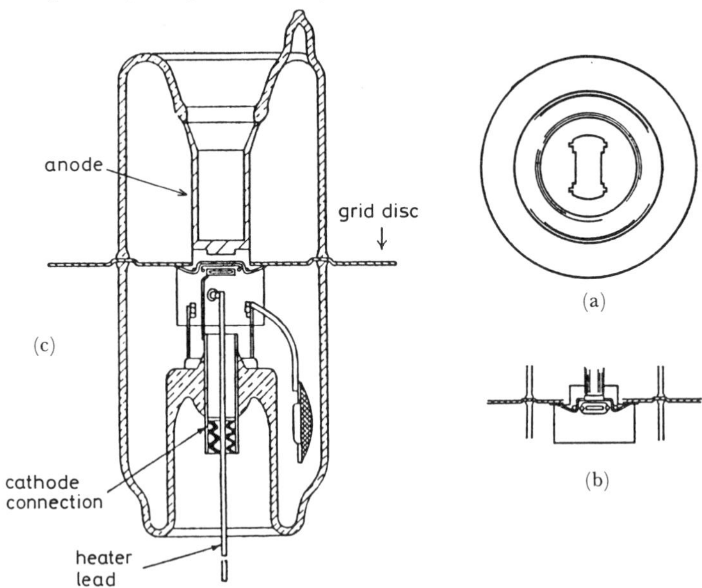

(a) Grid disc of the CV16, CV53 and CV88 showing the perforation for the pre-mount.
(b) Diagram showing the relationship between the pre-mount and the disc in the CV16, CV53, and CV88.
(c) Section through CV88 valve.

All of the valves in the series used the same grid-cathode assembly, in which the electrodes were accurately located by a pair of vertical mica insulators mounted on either side of a 'stem' which included the cathode and heater lead seals and formed part of the lower half of the envelope. To take advantage of established technology, the cathode was a standard rectangular nickel tube, coated only on the side facing the grid, which was wound and stretched in the usual way. To

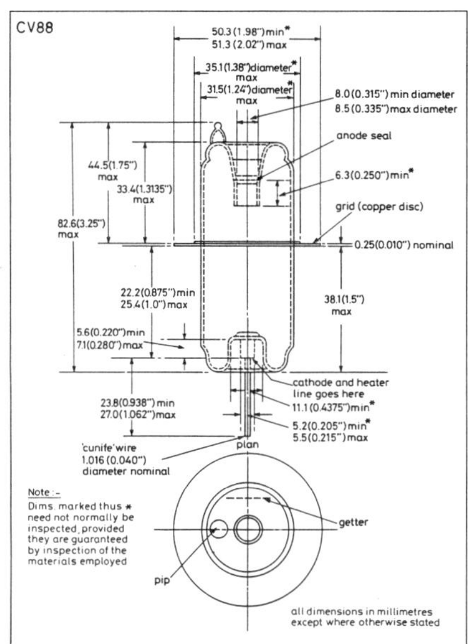

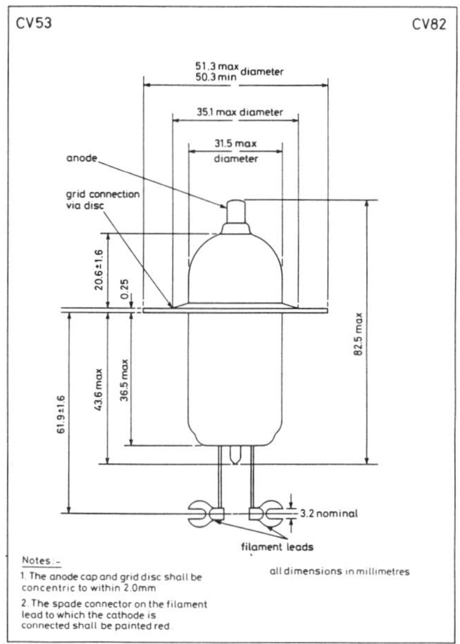

reduce the input capacitance, most of the grid wires facing the plain side of the cathode were removed, leaving a small number at each end to maintain the rigidity of the grid assembly. Four short wire spring contacts attached to the grid rods made eight contacts with the grid disc when the completed stem assembly was inserted into the envelope and sealed in position. The grid wires were 0.05 mm in diameter with a pitch of 0.165 mm. The anode-grid separation was approximately 0.6 mm, and the active area of the cathode about 6 mm by 3 mm. With 6.5 mA anode current, the transconductance was 6.5 mA/V and the amplification factor 100. The input impedance (about 150 ohms) was of the same order as the characteristic impedance of the twin wire and coaxial transmission lines with which it was used.

S25A, later known as CV16, was available early in 1941. Its anode, like that of a later variant, S28A (CV88), was a copper tube sealed into the upper half of the envelope and located by clips attached to the mica insulators supporting the cathode and grid. In practice, the bandwidth of the 600 MHz receivers (4 MHz) limited the external anode load to about 15,000 ohms, so reducing the gain to 16 dB (about half the amplification factor). Placed in front of the best receiver then available, which used a diode mixer as the input stage, it reduced the overall noise factor by about 8 dB. The cathode connection in CV88 was a copper tube with a concentric heater lead. This reduced RF losses in the cathode circuit, which could then be a coaxial line. This resulted in a further improvement in noise factor by about 4 dB. CV88 could be used up to 1 GHz.

After the tests at 600 MHz, it was found that S25A as a pre-amplifier gave an improvement of 4 dB in the noise factor of 200 MHz receivers using an RL7 pentode as the input amplifier. As a consequence, Foster and his team modified the design of S25A to use a nickel box anode with a wire connection to a cap at the top of the envelope. In addition to being easier and cheaper to fabricate and assemble than S25A, S26A (later known as CV53) was more suitable for use with tuned circuits using inductors and capacitors. It gave the same improvement in the noise figure of 200 MHz receivers, and could be used up to 400 MHz.

3.4.4 Signal Amplifiers above 600 MHz

After completing development of the 600 MHz series, Smyth and Foster investigated the possibility of extending the frequency of operation of grounded grid triodes by reduction of the grid-cathode spacing to the practical minimum. Using electroplated grids with 300 lines per inch and a grid-cathode spacing of 0.002″, they built an experimental valve with a cathode 1 mm in diameter, and an envelope with disc seals for the anode, grid and cathode leads. Its overall length was $1\frac{1}{2}″$, and its diameter $\frac{1}{2}″$. With 10 mA anode current, its transconductance was 5 mA/V. At 600 MHz the gain was 12 dB, falling to 6 dB at 3.3 GHz and zero at 4.7 GHz. The noise factor improvement over a crystal mixer was 6 dB at 600 MHz, 2 dB at 1.2 GHz and zero at 1.7 GHz. They concluded that without some new principle of operation and design, their figures could not be improved substantially. Their prediction was confirmed by tests made in 1941 by GEC on their E1411 (CV153), an amplifier version of the CV90 local oscillator triode which had disc seals for the anode and grid and a cathode lead forming part of a coaxial line. With amplification factor 80 and transconductance 7 mA/V at 10 mA anode current CV153 gave noise factors of 7 dB at 600 MHz, 9.5 dB at 1 GHz and 12 dB at 1.5 GHz.

3.4.5 Local Oscillators

At frequencies up to 200 MHz, local oscillators were basically 'general purpose' valves. Being self-oscillators, they did not require high gain to operate with acceptable efficiency. At 200 MHz, the first British radar receivers used acorn triodes, but these were superseded in 1940 by the Mullard RL18, which could generate up to one watt output at 25% efficiency and operate effectively with a small tuning range at frequencies up to 800 MHz.

The first local oscillator designed specifically for use at 600 MHz was the STC disc seal triode S27A, later known as CV82, a variant of the CV53 with an amplification factor of 35 and internal feedback between the anode and grid circuits. Available in 1941, it generated up to 1.5 watt output at 600 MHz with an efficiency of 25%, and could be tuned continuously from 400 to 750 MHz by adjustment of the anode circuit.

Similar performance was obtained by GEC from their CV90 triode, which was available in 1943. Designed by G.W. Warren, it used disc seals for the grid and anode and could be plugged into a double concentric coaxial line. Operating at frequencies up to 3 GHz, it generated up to 4 watts output at 1 GHz with an efficiency around 25%. In 1945, it was replaced by CV273, which used a similar but smaller structure and could operate up to 3.7 GHz.

In CV90, the oxide coated cathode was a short cylindrical cup with a flat top 4 mm in diameter, closed at the other end by a welded on thermal baffle after the

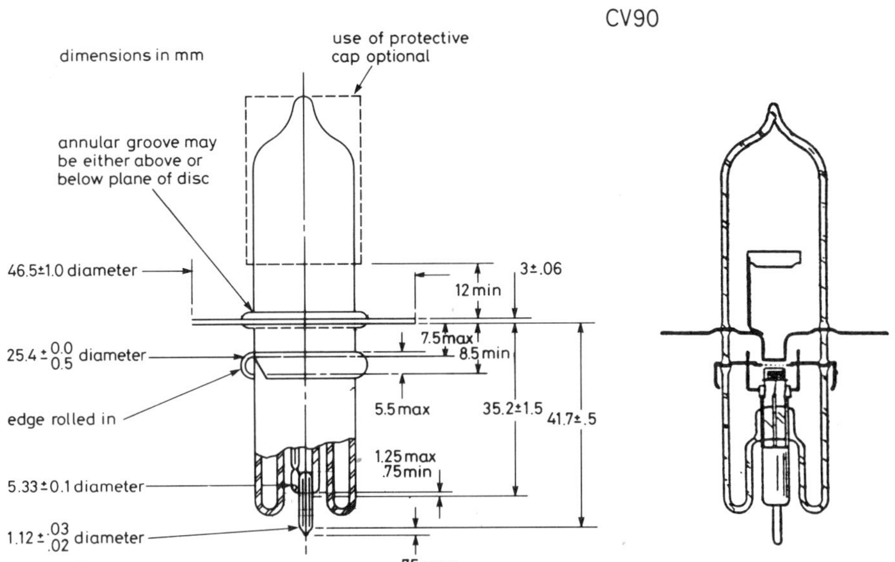

heater was inserted. The grids were wound with molybdenum wire 0.03 mm in diameter with a pitch of 0.145 mm across a rectangular molybdenum plate with a circular central aperture 5 mm in diameter. After winding two grids back to back, the edges of the plates were copper plated and the wires brazed in position in a hydrogen atmosphere, after which the grids were separated and tensioned by compressing the discs across a diameter at right angles to the wires. The grid and cathode were then assembled on a stem with a pinch housing the cathode support rods and the seals for the cathode and heater leads. The stem also carried two short support rods for the grid, sealed into a pair of glass pillars, one on each side of the pinch. The grid-cathode spacing was set by illuminating the assembly by oblique lighting from above, and observing when the shadow of one wire on the cathode surface was eclipsed by the adjacent wire. After this adjustment, which was made by compressing the support rods, the assembly was inserted into the anode-grid seal assembly and sealed in position with the grid in close contact with its disc, which had two locating holes for the grid plate supports on opposite sides of its central aperture. Strips of tin were attached to the support rods, which melted when the valve was baked on the pump, so making good thermal and electrical contact between the mating parts. The grid-anode spacing was adjusted by distorting the copper anode and grid discs after the seal sub-assembly was completed and before the grid-cathode unit was inserted.

Because of the high operating current density (almost 1 A per cm^2), special

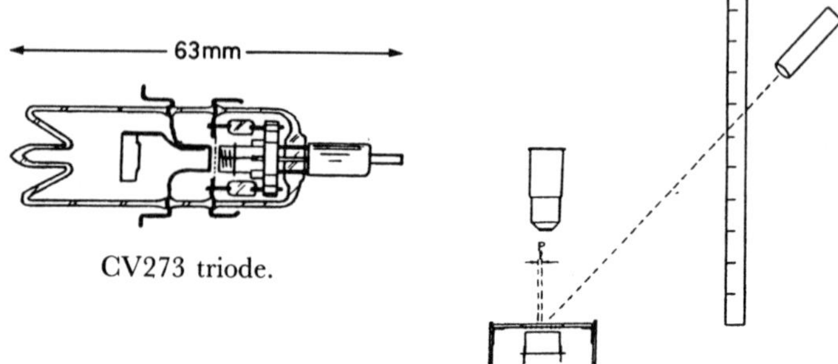

CV273 triode.

Use of shadowgraph for measuring grid-cathode clearance.

pumping schedules had to be developed to minimise contamination of the grid by a mixture of barium oxide and barium evaporated from the cathode during the exhaust process and thermal activation of the cathode surface. [Electron emission from the evaporated layer can be reduced by gold plating the grids, but this may produce nodules on the grid wires which affect performance adversely]

3.4.6 Mixers

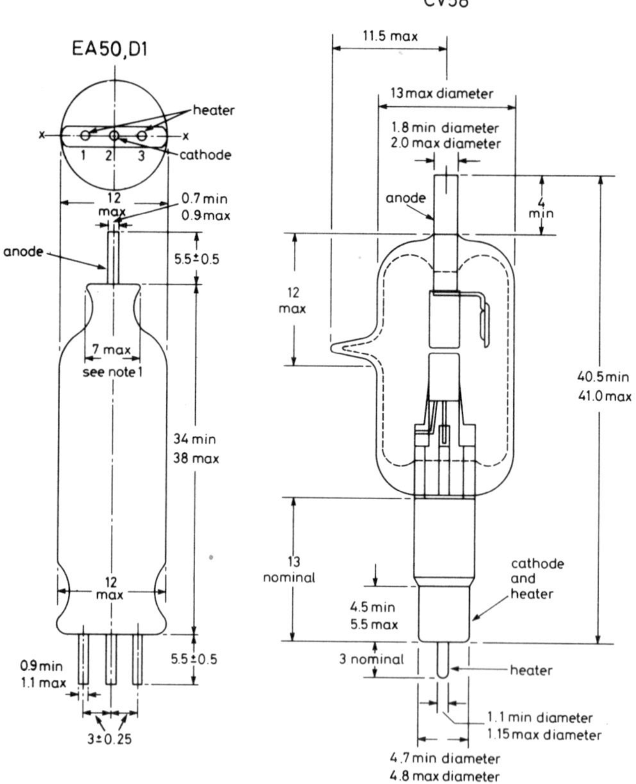

dimensions in mm

Because they used a signal amplifier as their input stage, the receivers in early radar systems such as CH and Type 79 could use 'television diodes' as mixers.

The performance of diodes as an input stage was acceptable at 200 MHz, but inferior to that of acorn signal amplifiers and significantly lower than that of the RL7 low noise pentode. Above 300 MHz, their performance deteriorated, becoming unacceptable at frequencies approaching 1 GHz. Crystal mixers gave significantly better noise factors at 600 MHz, but were much more vulnerable to damage by leakage power from their associated transmitters or by radiated signals from others nearby.

To overcome this problem, and to determine the feasibility of using diodes at microwave frequencies, GEC developed E1273, a diode suitable for use in twin wire or coaxial transmission lines. Available as CV58 in 1942, its anode-cathode spacing (0.05 mm) was the closest that could be achieved without risk of contaminating the anode by evaporation of the cathode coating or misalignment of the two electrodes. Its performance was comparable with but inferior to that of crystal mixers up to 1 GHz, and deteriorated rapidly thereafter.

3.5 Duplexers – TR and TB Switches

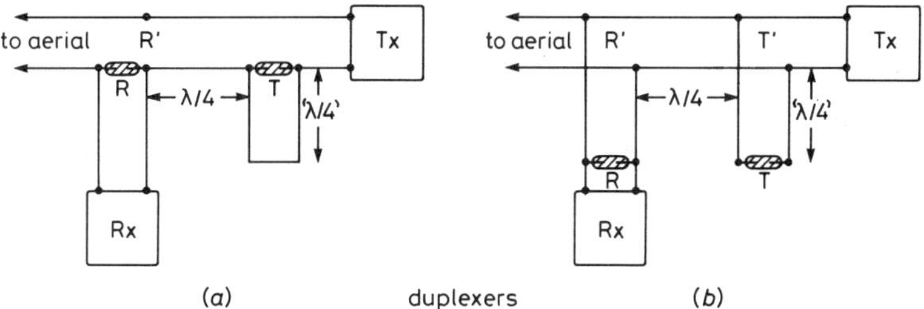

3.5.1 Foreword

In a 'common T and R' pulsed radar system, the duplexer has three functions:-
(a) to minimise attenuation of the radiated pulse when the transmitter is turned on,
(b) to provide adequate protection for the receiver during the pulse, and
(c) after the pulse has passed, to cause minimum loss of input signals generated by energy reflected from objects within the target area.

At frequencies below one gigahertz, receivers in wartime radar equipment used valves as signal amplifiers or mixers in their first stage, and so required much less protection from the transmitter than the crystal mixers in microwave systems. This made possible the design of duplexers in which the switching elements were either gas discharge tubes (spark gaps) turned on by the RF pulse from the transmitter, or vacuum diodes with the ability to pass high current in short pulses.

The simplest arrangement of a metric duplexer uses twin-wire transmission lines as shown at (a), in which the switches are connected in series with one of the conductors. Alternatively, they may be in parallel, as shown at (b); (an equivalent configuration could be used with coaxial lines.) In configuration (a) with the switches at R and T turned on, the RF pulse reaches the aerial with very little loss. At the same time, the low impedance at R limits the power breaking through to the receiver. With the switches turned 'off' after the pulse has passed, there is a high impedance at T which prevents all but a small fraction of the returning signals from entering the transmitter. [The 'quarter wave' line is actually shorter than one quarter-wavelength, and acts as an inductance in parallel with the capacitance of the switch. Thus the combination behaves as a true quarter wave transformer, converting the short at its end to a high impedance at T. (The intrinsic losses in transmission lines prevent the transformation of a 'short' to an 'open' circuit)]. At the same time, the high impedance at T is transformed into a low impedance at R, so that returning signals enter the receiver with very little attenuation. Because the input impedance of the receiver matches that of the line, the shunting effect of the switch capacitance does not cause a significant loss of signal. In configuration (b) the difference between the 'on' and 'off' impedances of the switches produces the same effect on the outgoing pulse and the incoming signals. To ensure that there is minimum loss of signals indicating targets close to the aerial, the switches must return to the 'off' condition within a short time after the pulse has passed. (A recovery time greater than ten microseconds will effectively prevent detection of targets at distances of one mile). When diodes are used as switches, the recovery time is intrinsically short. With gas filled devices, the recovery time is controlled by persistence of ionisation after the pulse has passed. This may require that the rare gas filling include a deionising agent such as water vapour or oxygen.

3.5.2 Spark Gap Switches

The first experiments with metric 'common T and R' systems were made at TRE in mid 1940. Used in the 1.5 m (200 MHz) CHL transmitter, spark gaps in air operated satisfactorily for a few hours with transmitter powers up to 25 kW peak, life being limited by burn back of the tungsten electrodes. This led to the development of VI507 by N.L. Harris and his group at GEC. VI507 was a sealed-off unit with tungsten tipped electrodes with a gap of 0.4 mm surrounded by a loose glass sheath which confined the glow discharge. Filled with argon at a pressure of 400 mm Hg and a droplet of mercury, it operated satisfactorily for several hundred hours at 25 kW peak. The recovery time was about 100 microseconds. Type Approved as CV86 in April 1941, it was being produced in quantity by MOV within a few months.

The introduction of CHL and GCI transmitters with powers up to 120 kW peak, and the demand for greater uniformity of characteristics, necessitated changes in the geometry of the electrodes and their sleeves. In the Mk2 version, each of the tungsten tipped electrodes was covered by a loose ceramic sleeve and a sealed-on glass bead, which was ground away to expose the flat end of the tungsten rod. Filled with argon at 600 mm Hg, VI507C gave lives around 200 hours at 120 kW. The recovery time was approximately 50 microseconds. Adding 5% of oxygen reduced this to about 10 microseconds at the expense of an acceptable increase in breakthrough of transmitter power into the receiver. [In parallel with the work at GEC, G. Fertel of Signal School experimented with water vapour as a deionising agent. Tubes with the argon-oxygen mixture were easier to produce with adequate life.] Development was completed in September 1941. By May 1942, production of the Mk2 CV86 at MOV exceeded 1000 per

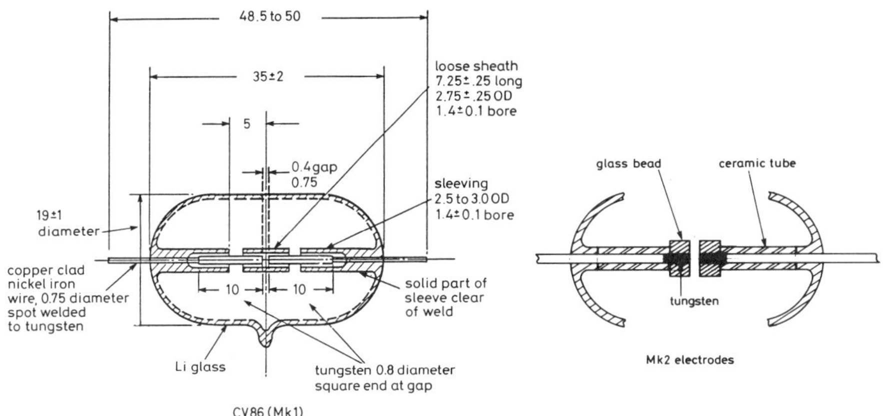

dimensions in mm

week. In addition to its CHL and GCI applications, CV86 was used in the Naval Type 291 air and surface warning radar, and in the 50 cm 100 kW transmitter which formed part of the AMES 11 mobile ground radar station developed at TRE in late 1941/early 1942.

3.5.3 Diode Switches

Unlike gas-filled devices, diode duplexer switches have intrinsically short turn-on and recovery times. Thus there is much less leakage of transmitter power into the receiver at the beginning of the pulse, and no problem with loss of signals indicating targets close to the aerial system. Requiring no water vapour or other deionising agent, they have inherently longer lives. Set against these advantages are the higher voltage drop across diodes when they are conducting and the need to arrange that the RF circuits in the duplexer accommodate the AC and DC connections to the heater and cathode and that the diode rectified current generate a bias sufficient to ensure no 'clipping' of large signals during the 'off' period.

In late 1940, soon after it became clear that the demand for two masthead positions for the Type 79 aerial systems was causing major difficulties with the installation of aerials for DF and communications equipment, experiments at Signal School indicated that diode switches could operate successfully in duplexers at frequencies up to 200 MHz, and so make it possible to use 'common T and R' arrays for early warning and search radar in capital ships and small craft.

The first installation to use a diode switched duplexer was Naval Type 279B where the transmitted output was 70 kW peak at 45 MHz. Because of the urgent need to release masthead space, there was insufficient time to develop a diode specifically for this purpose, and so each switch was an assembly of six television signal rectifying diodes connected in parallel to the coaxial line duplexer network. The first prototype system was designed by H.J. Reeves of Signal School and tested successfully at sea in April 1941. Very soon thereafter, units were in production at Signal School and fitted to all subsequent installations.

Paralleled diodes were also used as the TR switch in the coaxial line duplexer of Type 281B, operating at 90 MHz. Because the transmitter output could be as high as one megawatt in pulses with five microseconds duration, the duplexer included a spark gap pre-TR switch to reduce the power reaching the diodes. This used a pair of tungsten electrodes operating in air. Adequate operating life was obtained by daily adjustment of the gap to compensate for burn back of the tips. Development of the duplexer by Signal School began in early 1941, at the time when the Type 279B system was being tested at sea. By the end of 1941, the first installations were in operational use.

In parallel with the work at lower frequencies, Signal School experimented with diode switches at frequencies around 214 MHz, where GEC had already developed the Mk2 version of the CV86 spark gap switch for TRE. The intention was to take advantage of the longer life of vacuum diodes and use them in duplexers for small craft radar systems. As part of this programme, GEC began development of the E1248 coaxial diode in late 1940. This used an anode and oxide cathode assembly similar to that of a 'micropup' transmitter valve, with the cathode supported by a tungsten rod sealed through a dome attached to one end of the anode by a 'Housekeeper' seal. The power supply connections to the heater and cathode were made through a seal assembly at the other end. In use, the anode was clamped to the outer conductor of the coaxial line. The beaded end of the cathode support provided a capacitive contact to the inner so that the

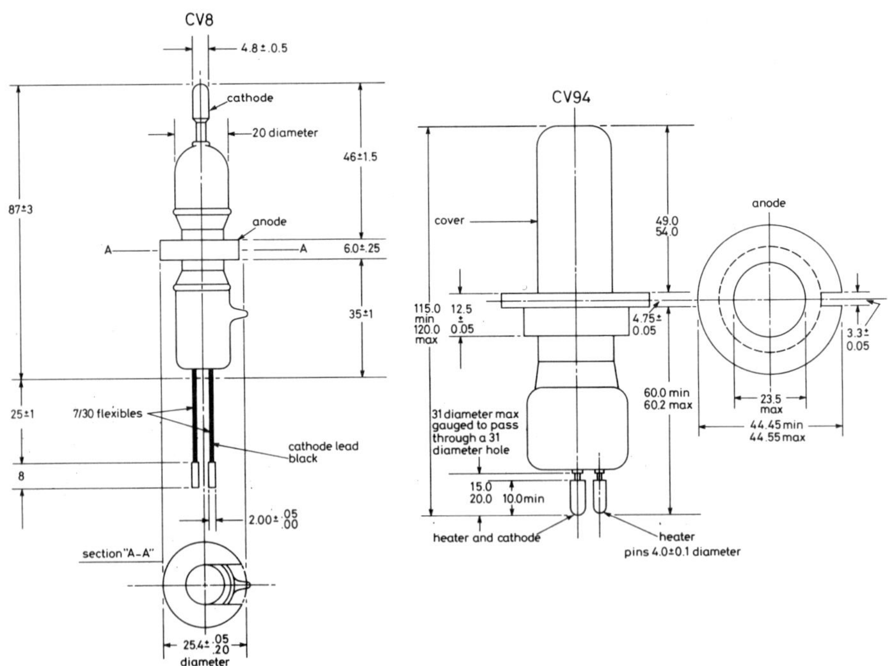

dimensions in mm

cut-off bias could be generated across a paralleled capacity and resistor connected in series with an isolating inductance in the lead from the other cathode connection to earth. Samples were tested by Signal School in April 1941, but after a short year's development, which included an increase in the dimensions of the active geometry, attention was concentrated on the development of a gas filled variant using the same anode and cathode (uncoated and minus its heater assembly). Filled with argon at 30 mm Hg and a small percentage of water vapour to reduce the recovery time to a few microseconds, E1421 operated successfully with inputs of 100 kW peak. By attaching a capsule containing 50 mgm of silica gel saturated with distilled water to the redundant heater connections inside the bulb, lives of several thousand hours were obtained (many times that of the Mk 2 CV86). From early 1942, E1421 was used in the Type 286 small craft radar, operating at 7 kW peak. In mid-1943 it was fitted to its successor, Type 291, operating at 100 kW. Postscript: Although E1248 was Type Approved as CV8 in 1942, no records have been found of its use in operational equipment. In 1943, Marconi's Wireless Telegraph Company was given complete responsibility for development of the Type 960 high power radar, which replaced both Type 279B and Type 281B in capital ships. Transmitting up to 450 kW peak at 90 MHz in pulses of 5 or 15 microseconds duration at a repetition rate of 250 pulses per second, its mean power output was several times that of Type 281. To eliminate the spark gap pre-TR, MWT developed a coaxial duplexer system using DS103, a vacuum diode switch which they also designed and manufactured. This had over two times the active area of CV8 and many times that of the paralleled diodes used in Type 281, with a correspondingly lower impedance which enhanced its ability to protect the receiver. Because of its greater interelectrode spacing, it was also less vulnerable to damage by the transmitter pulse. In 1945 it was Type Approved as CV94.

Cm Research 1940

We worked all through those early months, fighting,
first to be allowed to work at all,
and then for space and light and heat to work in;
for we were thought fit but for the next war,
or next year's war at best, and next year
was like a mirage in the desert sky,
when June had drained the last drops of complacency,
and France had written finis to that phase.
And so alone,
we, fighting every inch of the way,
against those ingrained elephants of inertia,
against the prejudice and the pride
of self-established, self-supporting systems,
we fought (through forests, thick with self-satisfaction)
to shorter electromagnetic wavelengths.

H.W.B.S.(1941?)

'Cm Research 1940' was written by Dr. Skinner in the year after the fall of France and the evacuation from Dunquerque in the Summer of 1940. It was found during a search of the TRE archives in October 1988.
E.B.C.

Chapter 4

Microwave Devices

4.1 A summary of achievement

Demonstration of the first British multicavity magnetron by J.T. Randall and H.A.H. Boot in February 1940 was the most important factor in stimulating development of microwave radar, both in Britain and the USA. Evolution of the new technology depended equally upon the development of broad band receivers with microvolt sensitivity and of magnetron transmitters giving hundreds of times the peak output of the Birmingham experimental model. In many ways, the design of active devices for the front end of the receiver was much the more difficult problem.

> The radar equation (see Glossary) shows that when 10 kW peak at 10 cm is radiated from a typical escort vessel aerial system, the power returned after reflection from a conning tower or a small aircraft at 3 km range is theoretically between 10^{-7} and 10^{-8} watts. At 30 km, the return from a large aircraft is about three orders less. With a receiver capable of detecting 10^{-12} watts, the theoretical maximum range for a bomber is around 60 km.
>
> Taking account of losses in the TR cell, which may be around 2 dB, 10^{-12} watts is equivalent to an input of about ten microvolts to the receiver.
>
> Although receivers designed in 1944 could operate effectively with inputs around 10^{-13} watts, those available in 1941 required at least ten times that amount. Thus detection of aircraft at ranges over 50 km required development of transmitters generating peak powers of the order tens to hundreds of kilowatts at 10 cm.

In mid-1940, when the first high power magnetrons became available for use in experimental radar systems, there was no scientific basis for the design of low noise high gain microwave signal amplifiers. (The travelling wave tube was several years away). Thus the first stage of the receiver had to be a mixer operating with no more current than that required for minimum frequency conversion loss. [Simple rectifiers cannot operate at the microvolt level]. It also had to be reasonably immune from transit time effects. No thermionic device could meet these criteria. [Because of transit time effects and shot noise in its anode current]. Thus H.W.B. Skinner's development of the 'cat's whisker' crystal into the functional prototype of a practical microwave mixer by July 1940 was an important contribution to the new technology (see 4.4).

[The proposal that point contact crystal rectifiers be used as microwave mixers came from M.L. Oliphant, who made a small number of experimental units. In the 1920s, 'catswhisker' crystal detectors with galena (crystalline lead sulphide) as the active element were used in radio receivers with headphones.]

Irrespective of the availability of the magnetron*, the evolution of microwave radar depended upon R.W.Sutton's deciding in early 1940 to make a single resonator klystron oscillator, and to use a reflector electrode to initiate the second transit of the electron beam through the interaction space. The NR89, developed by his group at Signal School, and available for experimental use in the autumn of 1940, was the precursor of a family of tunable local oscillators with adequately low noise output and the ability to operate with automatic frequency control. [Because of the very low signal input to the mixer, noise injected by the local oscillator can seriously degrade the overall noise factor of the receiver. When klystrons are used their beam current must be as low as possible. Low power magnetrons cannot be used as local oscillators because of the inherently high noise content of their output signal].

*Experimental high power klystrons intended for use in radar transmitters were being made by M.L. Oliphant's group at Birmingham University when the first British multicavity magnetron was assembled there by Randall and Boot in 1940. The performance of the magnetron soon surpassed that of the VFO ('Valve for Oliphant') and so British development of high power klystrons ceased in 1940. It remained in abeyance until after the invention of the multiresonator very high voltage klystron amplifier by Ginzton and the Varian brothers at Stanford University in 1947. Attempts were made at Birmingham to use the same klystron as a transmitter and a local oscillator by changing the operating conditions. These were unsuccessful, partly because of the high beam current in the low power mode of operation.

The TR cells invented by A.H. Cooke of the Clarendon Laboratory in 1941, and described at that time as 'soft Sutton' tubes, were not essential to the development of the first microwave radar systems. (The Army and the Navy used twin mirrors in fire control and gun-laying equipment developed in 1941.) They were, however, essential components in aircraft radar such as AI and H_2S.

The ingenuity, and the science and technology applied to the family of receiver components was in every sense comparable with that applied to the magnetron. Perhaps the most important characteristic of the evolution of the whole family of active devices for microwave radar was that it came about as a result of equally significant innovations by a number of workers in the field.

After Randall and Boot's demonstration of their 6-cavity 10 cm magnetron, which was fitted with a small diameter tungsten filament cathode and therefore basically a device for continuous operation, Megaw and his group at GEC made essential and fundamental changes to the original Birmingham design, and in mid-1940 developed E1189, the first magnetron suitable for use as a radar transmitter*. This had a large diameter oxide cathode with end discs, similar to one used earlier by H. Gutton of SFR in his 8- and 12-segment interdigital magnetrons. Some years previously, Megaw, who was in close contact with Gutton, had identified back bombardment of the cathode and the importance of secondary emission as the major source of operating current. He had also ventured the opinion that magnetrons would operate efficiently with cathodes significantly greater in diameter than the tungsten filaments and small spirals in common use at that time. It was Gutton who demonstrated the manyfold increase in power which could be obtained by replacing a tungsten spiral by a large oxide cathode, and that cathodes of this type could withstand operation with very high voltages. Megaw himself acknowledges the great importance of Gutton's innovation, which was the key to development of multicavity magnetrons giving peak power outputs from ten to several hundred kilowatts. Without the invention of 'strapping' by J. Sayers of Birmingham University and the refinement of this technique by S. Devons of TRE , who originated 'strap setting', the multicavity magnetron would have continued to be inherently unstable in frequency and difficult to operate. The addition of straps also increased the efficiency and power output by many times. Without them, development of the 'megawatt' S-band magnetron would have been delayed by many years. The addition of straps also increased productivity in manufacture. Later on, B.V. Rollin of the Clarendon Laboratory, who had concentrated on designs for wavelengths down to 1.25 cm, invented the 'rising sun' resonator system. This required no straps, and so was much easier to fabricate than the ring-strapped system. (The rising sun design was also used by Columbia University, New York in their X- and K-band magnetrons. It is almost certain that they originated their own design independently)

* In a letter to Dr. Megaw's secretary shortly after his death in 1956, Sir Edward Appleton, who in 1940 was a member of the Committee for the Scientific Survey of Air Defence (the 'Tizard' Committee), wrote 'Those who were in the business know how much the practical development of the cavity magnetron – the development that made it something that could go into operational use – was due to Dr. Megaw.' [The Tizard Committee reported to and made recommendations for action by the Subcommittee on Air Defence Research which reported to the Committee of Imperial Defence.]

In a similar way, the evolution of the original 10 cm 'Sutton tube' into a family of stable and easily used microwave local oscillators came about as a result of a theoretical study of its important characteristics by L.F. Broadway and his group at EMI, followed by their making significant changes in design, and developing sophisticated technology for their manufacture. There were also important contributions from J. Daunt and D. Roaf of the Clarendon laboratory, Oxford, particularly in extending the operating range down to 1.25 cm.

After H.W.B. Skinner's demonstration that the silicon 'cat's whisker' crystal operated effectively as an S-band mixer, T.H. Kinman of BTH developed the CV101 ceramic capsule assembly which was easy to mount in waveguide systems and robust enough to withstand vibration in aircraft. The capsule later became the international standard package. A study of the metallurgy of the 'doped' silicon active element by J.W. Ryde of GEC resulted in the development of mixers less liable to damage by the minute fraction of transmitter power passing through the TR cell.

Following A.H. Cooke's original concept of the S-band 'soft Sutton' tube, H.W.B. Skinner and A.G. Ward of TRE and A.R. Chance of E.K. Cole continued its development as tubes were manufactured for Service use. This resulted in the evolution of TR cells with a keep-alive electrode replacing the radioactive material used initially as the source of primary ionisation. [To ensure rapid build up of ionisation by the transmitter pulse, there must be some ions in the rhumbatron gap between pulses but not so many as to cause attenuation of incoming signals]. After Skinner and Ward had followed this by making an experimental 3 cm TR cell, N.L. Harris and his group at GEC developed new designs and fabrication technology for X-band TR and TB cells.

It is worthy of note that at the end of 1939, when the decision was made to develop 10 cm radar, there were no suitable active devices available – no high power magnetron, no reflex klystron, no proven microwave crystal mixer and no TR cell. By mid 1941, Type 271, the first Naval S-band radar, was in operational use. The first operational model of the GL3 S-band gun-laying equipment for the Army had been delivered to ADRDE. AI MkVII, the first centimetric airborne radar, was being flown experimentally, and used in combat in December.

The 'Trigatron' modulator switches developed in 1942 by J.D. Craggs, M.E. Haine and J.M. Meek of MetVick were essential to the full exploitation of magnetron technology in airborne radar. When E1198/CV38 was fitted in AIMkVII in late 1941, its specified maximum peak input power was 100 kW, roughly equal to that available from hard valve modulators small enough and rugged enough to be used in aircraft equipment. The trigatron CV85 in Modulator 64 generated up to 170 kW, so that CV64, the strapped successor to CV38, gave outputs around 40 kW in AIMkVIII and $H_2 S$. Within about a year, CV125 in Modulator 158 gave up to 500 kW input to CV192, an uprated version of CV64, so raising the transmitted power output to around 200 kW, about twenty times that of CV38. Being cold cathode devices, the trigatrons required no heating power, which simplified equipment design and cooling. [Trigatrons were also used in mobile and aircraft decimetric radar systems.]

[Mercury thyratrons, which gave higher power output, could not be used in aircraft equipment because they were limited to pulse intervals greater than one millisecond by their deionisation time (see 6.5) and the decrease in vapour pressure resulting from the low ambient temperature at high operational altitudes. (It was not practicable to provide a controlled temperature enclosure for the valve.) Rotary spark gaps in air were also limited to operation at low altitude. Having no trigger system to initiate the discharge they were also prone to 'jitter', a random fluctuation in pulse interval, which made it difficult to design accurate timebase circuits in the display unit.]

The 'afterglow' phosphors developed in 1941/42 by teams led by L.C. Jesty and R. Puleston were essential to the development of the Plan Position Indicator. [It depended equally upon the use of aerial systems with the high directivity made possible by microwave techniques]. First envisaged by E.G. Bowen in 1936 it was one of the most important advances in military and civil radar tehnology.

Apart from those remembered for their innovations, there were others too numerous to mention who made valuable contributions to the development of new devices, or created new measurement and manufacturing techniques. The extremely rapid growth of microwave technology was a testimony to their collaborative effort within CVD, which quite often involved free exchange of 'know how' between organisations which were highly competitive in peace time.

4.2 Magnetrons

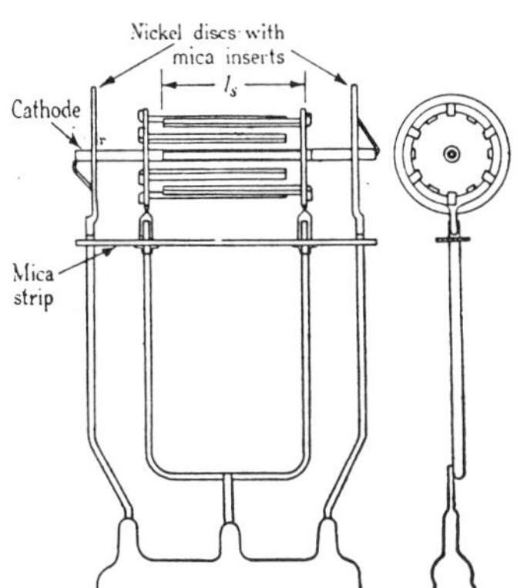

12-segment interdigital magnetron

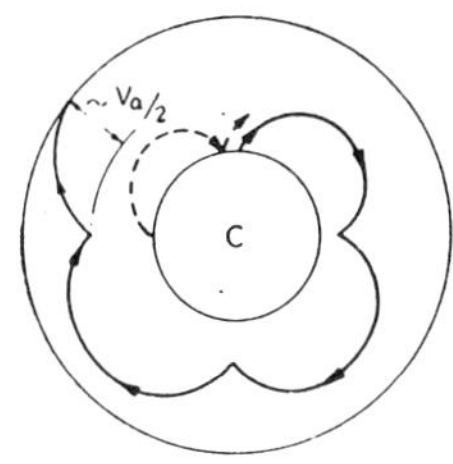
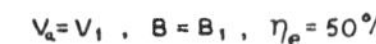
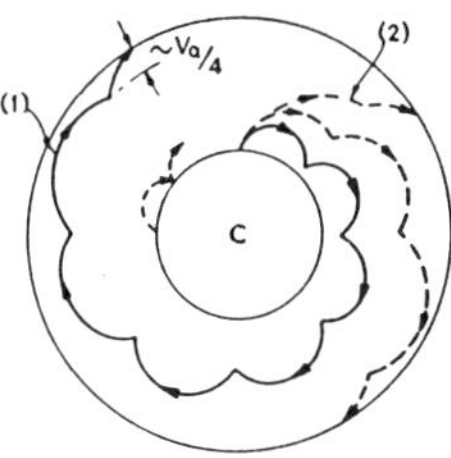

Electron orbits in an oscillating magnetron, showing effect of increased magnetic field and r.f. voltage. (1)—Trajectory with low segment voltage. (2)—Trajectory with high segment voltage.

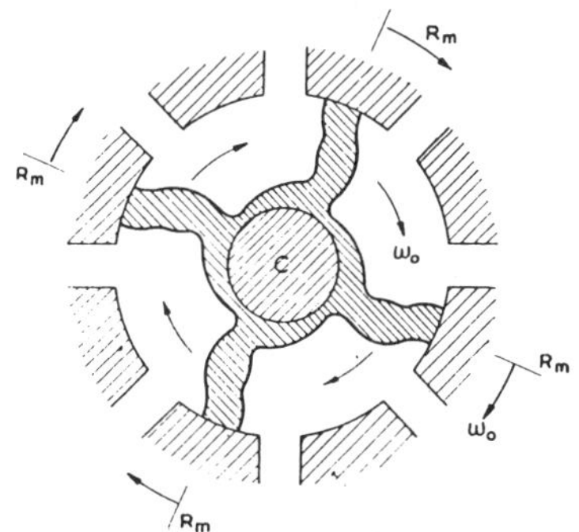

The rotating electron cloud in an 8-segment magnetron.
R_m—position of maximum retardation.
$$\omega_0 = \frac{2\pi f}{4}.$$

Magnetrons have a cylindrical anode with a cathode along its axis. Under the combined influence of a radial electric and an axial magnetic field, electrons leaving the cathode move towards the anode in cycloidal paths, interacting with the RF fields fringing longitudinal gaps in the anode wall. Some gain energy and return to the cathode. Others give up energy which can be coupled out of the system, and are collected by the anode as current which provides the input power. The gaps in the anode are part of the resonant system which determines the frequency of oscillation.

4.2.1 Early Days

From about 1930, Laboratories in Europe, USA and Japan were engaged in research on 'split-anode' magnetrons as sources of power at very short wavelengths, the intention being to use them in communications equipment.

In 1932, E.C.S. Megaw and a small group at GEC were studying propagation below 60 cm, and developing magnetrons and other components for use in measurement apparatus and experimental transmitters. In 1933, Megaw demonstrated that magnetron cathodes were heated by returning electrons, and predicted that 'secondary emission from the cathode (of a magnetron) may be a factor of considerable practical importance in the production of very short wavelengths.'

In 1934/5 K. Posthumus of Philips, Eindhoven successfully developed a 4-

segment magnetron. In March 1935, he published in the 'Wireless Engineer' a paper on his 'rotating electron cloud' theory of magnetron operation, which explained how 4-segment tubes could give a performance superior to that of 'split-anode' (2-segment) types, and how the use of a multiplicity of anode segments reduces the magnetic field and the voltage at which a given efficiency can be obtained. Thus it could be said that Posthumus' analysis leads directly to the multicavity designs which followed publication of his paper, and to Gutton's experiments with 8- and 12-segment interdigital magnetrons. It is recognised as one of the most important papers on magnetron theory. On the basis of Posthumus' hypothesis, it can be shown how some of the electrons precessing towards the anode gain energy from the RF field in the interaction space and return to the cathode, so heating it by 'back-bombardment', and generating secondary emission. The RF output power is generated by the greater number of electrons which give up energy in the interaction space and are collected by the anode. As a result of this energy exchange, the electrons are grouped into rotating 'spokes' which interact synchronously with the RF fields in anode-cathode space. In 1935, in a joint letter to Nature 1.6.1935,914, Posthumus and Megaw show that the 'rotating electron cloud' mode of operation of 4-segment types observed by Posthumus is the same as one mode of operation of split anode magnetrons*, and that in both cases the energy of electrons reaching the anode is less than that corresponding to the anode voltage.

[*see also 'An experimental study of low power CW magnetrons having few segments'. E.B. Callick, Proc. IRE Vol. 40. No. 7. July 1953]

In 1936 Megaw confirmed his earlier experiments on back-bombardment of magnetron cathodes, and concluded that secondary emission could provide a major fraction of the operating current. He also ventured the opinion that magnetrons could operate efficiently with cathodes considerably larger in diameter than those in common use at that time.

In January 1938, in papers published in 'Journal de Physique' and the 'SFR Bulletin', H. Gutton and S. Berline of SFR in Paris described 8- and 12-segment interdigital magnetrons with tungsten filament cathodes. Operating at 700 V, the power output was ten watts CW at 16 cm wavelength and 6 W at 10 cm. In June 1939, Megaw, who was a personal friend of Gutton and Dr. Maurice Ponte of CGTSF, visited the SFR laboratory and was shown M16, a similar magnetron with 12-segments which gave 50 W peak at 16 cm when pulsed in an RDF system which detected ships at 10 km. He was promised samples, but their dispatch was delayed by the outbreak of war. During his visit, Megaw suggested that Gutton replace the tungsten filament by a thoriated tungsten spiral considerably greater in diameter. [It was thought at the time that only tungsten cathodes could withstand operation at high voltages]. The results of Gutton's tests and trials with the filamentary M16 were reported by Megaw at the CVD meeting in August 1939. At about this time Gutton decided to fit an oxide cathode with a diameter about 0.4 of that of the segmented anode. The immediate result was a tenfold increase in peak power output to 500 W, with secondary emission providing about 99% of the operating current. [This was the first practical demonstration of Megaw's prediction that magnetrons would operate efficiently with relatively large cathodes, and that secondary emission would play an important part in their operation.]

[The secondary emission ratio of tungsten is about 1.4 or less, and so secondary electrons cannot provide a major part of the operating current in magnetrons with tungsten filament cathodes. (The small diameter of the cathode is also a contributary factor.) This is consistent with reports that the power output of magnetrons with filament cathodes is limited by primary emission when attempts are made to pulse them above their CW design voltage and current. The secondary emission ratio of an oxide cathode is between 2 and 4 and so relatively few cycles of electrons returning to the cathode will generate many times the available primary emission. (The chosen larger diameter of the oxide cathode is also a contributory factor.)]

In December 1939, under Air Ministry contract, GEC started development of an AI system operating in the ultra short wave band, the target wavelength being 25 cm. The first experimental transmitters operated satisfactorily with E1130, a 'millimicropup' triode developed by GEC within the CVD contract. In March 1940, it was agreed that the target wavelength be changed to 10 cm. [This was the wavelength selected in 1939 by E.G. Bowen of Bawdsey Research Station and C.S. Wright, DSR Admiralty because operation at this wavelength would make possible the production of narrow beams by paraboloid mirrors small enough to be fitted in fighter aircraft. It was also the wavelength chosen by A.B. Wood of Signal School as being the best prospect for detecting small objects at sea level]. Megaw and his group then prepared designs for magnetrons giving up to 500 W peak at 10 cm, including a 4-segment glass envelope tube with a thoriated tungsten spiral cathode and an internal resonator system coupled into a waveguide.

4.2.2 The Multicavity Magnetron

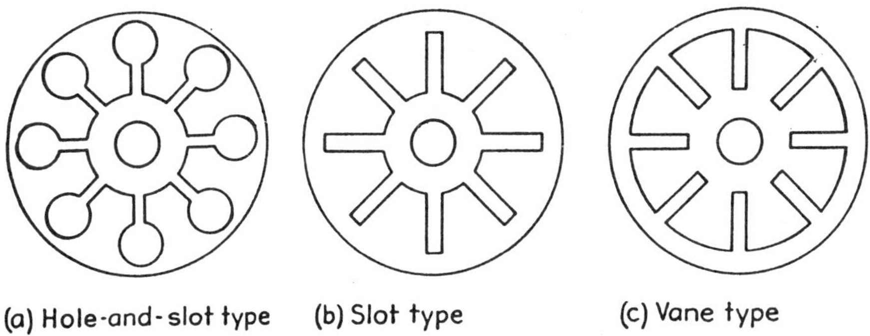

(a) Hole-and-slot type   (b) Slot type   (c) Vane type

In US patent No. 2063342, first filed in December 1934 and issued two years later, A.L. Samuel of the Bell Telephone Laboratories referred to magnetrons having 'a plurality of resonators', and described in detail a 4-cavity unit in which the anode block formed the centre section of a thin-walled copper cylinder. The thoriated tungsten filament cathode was supported by tungsten rods sealed into hemispherical glass domes which were themselves sealed to the ends of the copper cylinder. Output connections were provided for the attachment of a twin wire transmission line. The magnetic field was provided by an electromagnet surrounding the cylinder.

[In mid-1937, A.B. Wood of Signal School suggested in an internal memorandum that a similar structure, albeit in an all glass envelope, could be used as a radar transmiiter. He had previously concluded on theoretical grounds that for detection of small objects at sea level, systems operating at 10 cm would be preferable to metric radar designed primarily for the detection of aircraft. No action was taken on his proposal.]

The first British multicavity magnetron was conceived by J.T. Randall and H.A.H. Boot at Birmingham University in the Autumn of 1939, and operated for the first time on 21st February 1940. It was a demountable unit, having a water-cooled copper anode block with six 'hole-and-slot' resonators opening into a 12 mm cylindrical cavity containing the 0.75 mm tungsten filament cathode, which was supported by seal assemblies waxed to the end of the block. It differed from the Samuel magnetron in being designed to operate at a shorter wavelength, in having six resonators with narrower slots than those illustrated by Samuel, and a coaxial output line connected to a loop in one of the cavities. The magnetic field was provided by a large electromagnet with a five inch gap. Operating at 8kV, it was estimated that about 500 watts CW was obtained in short bursts with the tube continuously pumped. This was about ten times the highest power then known to have been obtained in the region of 10 cm. [Gutton's results with the oxide cathode were not known to British workers because of the ban on communications imposed by French security following the outbreak of war].

At the CVD meeting on 5th April 1940, W.B. Lewis of AMRE described the Birmingham magnetron and the results obtained on the pump. As a consequence of this, Megaw visited Randall and Boot on 10th April and suggested that their design could be improved and simplified by using side-arm supports for the filament and closing the ends of the block by soldered-on copper discs. This would make possible a great reduction in the gap of the magnet, and so of its weight. At first, the Birmingham group were reluctant to accept these modifications because they considered that the mode of operation of their tube was not fully understood, and therefore that it should be put into production without modifications which might cause delays. It was agreed that GEC would make sealed-off magnetrons for the Birmingham group and help in every way possible with techniques and materials. Randall and Boot would provide a drawing indicating the main features of the design. This arrangement was confirmed in correspondence between C.C. Paterson and M.L. Oliphant 18/20th April 1940, and reported a little later to DSR. The drawing sent by Randall and Boot early in May showed the original block and filament dimensions, and incorporated the filament seal assemblies and the end closures suggested by Megaw. On this basis, GEC prepared a complete design, E1188, in which the end spaces were reduced to the minimum so that the magnetron could be operated in an electromagnet with a 7 cm gap. The filament assembly was modified by the addition of tantalum endshields and the removal of the 'cool ends' from the interaction space. The blocks were machined in Birmingham. Tests on a sample tube taken to Birmingham in July showed that it reproduced the results obtained from the demountable unit. A few days later, a letter from Oliphant to Paterson refers to 'the great success of the modified form developed by Megaw'.

At the time of Megaw's first visit to Birmingham, GEC's interest in the Randall and Boot design was its possible use as a high power transmitter in their Admiralty USW communications project, but the increasingly urgent need for a high power 10 cm pulse transmitter for AI led them to the conclusion that a multicavity magnetron based on the Randall and Boot anode block design would be a much better prospect than their existing 4-segment experimental units. [Because the Birmingham design used a tungsten filament cathode, it and E1188 were basically CW magnetrons, limited in power output by the available primary emission.] Megaw and his group then prepared a design for E1189, a pulsed magnetron having an anode block with cross-sectional dimensions almost the same as E1188, except that the wall thickness between the resonators was increased to improve cooling of the anode, which was fitted with annular fins for air cooling. It was assumed that the mode of oscillation in the Birmingham tube was that described by Posthumus and therefore the same as that in Gutton's multisegment interdigital magnetrons, and that the wavelength of the resonators in the copper block was almost independent of their length. (Randall and Boot had also made this assumption). On the basis of design formulae worked out by Megaw and his group in 1939, the ratio of anode length to diameter was chosen to give the best compromise between output power and magnet gap. The cathode was a thoriated tungsten spiral with a diameter 0.4 of that of the anode. The copper end caps were attached to the block by a gold ring diffusion process invented by D.A. Boyland of GEC in 1935. The end spaces were made as small as possible. As a result, the tube was considerably shorter than E1188, and could be fitted into an existing 6 lb 1000 Oersted permanent magnet. .

On 9th May 1940, shortly after the design was completed, Dr. Maurice Ponte of CGTSF brought to Wembley samples of Gutton's M16 with an oxide cathode. (These were the tubes promised earlier and now released by the French government.) Tests at Wembley, which confirmed Gutton's results, were to have a profound effect on the design of high power magnetrons thereafter. They showed:-
(a) the dominant role of secondary emission from oxide cathodes, which could provide almost all the operating current.
(b) that oxide cathodes could withstand operation at much higher voltages and powers than had previously been   thought possible. [This was also to have an equally profound effect on the design of high power triodes and tetrodes used in short and ultra short wave communications transmitters and in pulse modulators.]

Because of the materials used in its construction, a copper block magnetron cannot be processed to achieve and maintain a very high vacuum. Thus survival of the oxide cathode depends upon its being protected from positive ion bombardment, This is provided by the electron sheath surrounding the cathode during operatioin.

The design of E1189 was therefore modified to use oxide cathodes similar to Gutton's design. Megaw himself acknowledges the great importance of Gutton's innovation, which opened the way to development of magnetrons giving pulse power outputs from ten to hundreds of kilowatts in the S-(10 cm) and X-(3 cm) bands.

At the CVD meeting on 17th July, a report was given on E1189 serial Nos 1

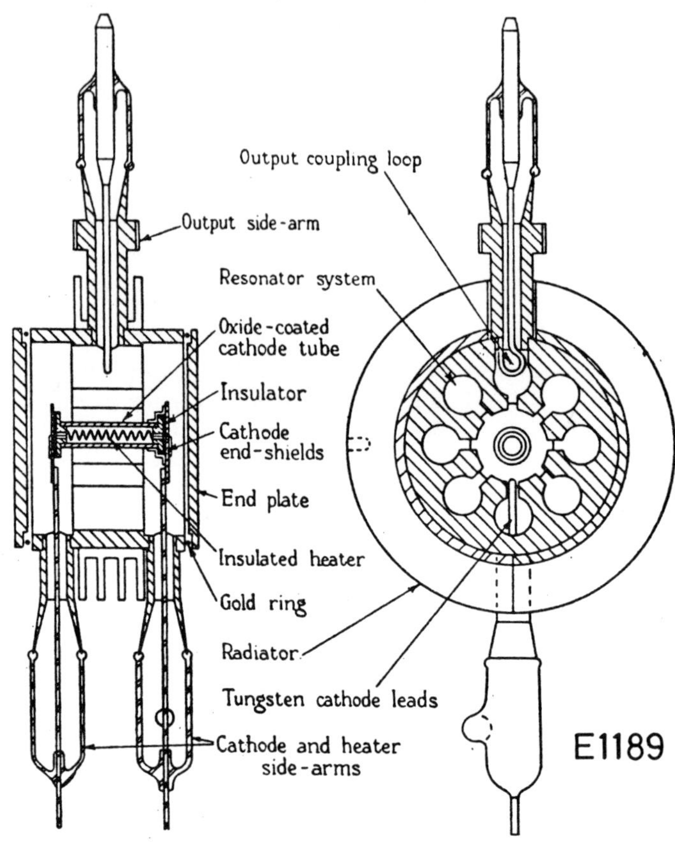

and 2. Operating at 9.5 cm, 5 kV, 1000 Oe, the peak output was 3 kW. At 8.5 kV, 1500 Oe they gave 12 kW, the field being provided by an electromagnet. The power output was independent of heater voltage. [On 24th July, serial No 4 was sent to Birmingham.] In July 1940, samples of E1189 were taken to TRE Swanage for tests by H.W.B. Skinner and J.R. Atkinson in their experimental radar system. In August 1940, a revised design of E1189 with 8-segments and a larger cathode gave approximately the same output when operated in the 6 lb 1000 Oe permanent magnet. [The increase in the number of segments and the consequent reduction in magnetic field is consistent with Posthumus' theory of magnetron operation.]

In August 1940 sample No. 12 of E1189, which had eight segments, was taken by the Tizard mission to the United States, and there copied by Western Electric, the manufacturing company associated with the Bell Telephone Laboratories, and by Raytheon who later worked very closely with the Radiation Laboratory of MIT (the Massachusetts Institute of Technology)

[After Pearl Harbour and the entry of the United States into the war at the end of 1941, 'BTL' and the 'Rad Lab' became two of the most important centres for research and development of radar systems and components for the allied Services].

E1189 was fitted in Naval radar Type 271, which was in operational use by July 1941, and later standardised as NT98. E1198, a variant operating at 9.1 cm, was used in early models of AI Mark VII in March 1941, and later standardised as CV38. In late 1940, BTH joined GEC in production of NT98 and CV38. By the end of 1941 the two companies had between them manufactured more than 2000 tubes. This was the beginning of a close collaboration in development and production which continued throughout the war.

4.2.3 Strapping

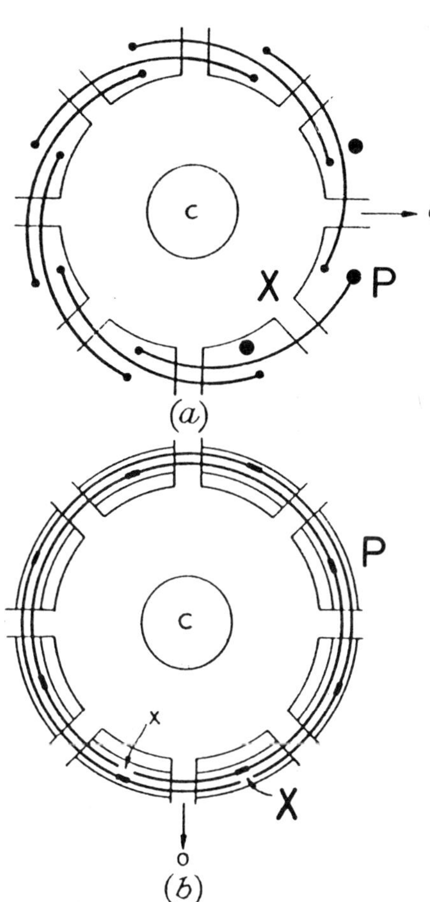

Strapped magnetron anodes.
O = output connection
P = points of connection of straps to segments
X = strap breaks
(a) Echelon straps
(b) Double-ring straps

X – In simple terms, strap breaks 'lock' the field pattern in the interaction space so that it is correctly oriented with the output connection, so ensuring optimum transfer of power from the block to the external circuit.

Because it has a number of resonators coupled together, a multicavity magnetron anode can support oscillations at more than one frequency. Thus interaction with the rotating electron cloud may generate power output at frequencies other than that of the preferred 'pi-mode', where the voltages on adjacent segments are 180°

out of phase. [It is for this reason that magnetrons have an even number of segments] This can result in 'moding', where the output jumps from one frequency to another nearby as the power input is increased from the operating minimum, and in the generation of spurious output at frequencies more remote from the wanted signal. The freedom to oscillate in more than one mode also limits the power input because it can result in collapse of the interaction associated with the pi-mode. The effects can be minimised to some extent by careful and often lengthy adjustment of the matching elements in the waveguide system to which the magnetron is connected (as, for example, the technique developed by the GL3 teams at BTH and ADEE during the early days of equipment design), but the basic problems remain.

Thus the invention of 'strapping' by J. Sayers of Birmingham University in July 1941 was a major advance in magnetron technology. In simple terms, strapping is the joining of alternate segments of the block by thick copper wires at each end so that unwanted modes of oscillation are suppressed. Thus the electrons interact only with the pi-mode, mode jumping is minimised if not eliminated, and the efficiency of operation is increased. This was Sayers' original premise. Subsequent analysis showed that the effect of strapping is much more complex, but the practical effect was immediate. [The principal effect of strapping is to shift the frequencies of unwanted modes away from that of the pi-mode. In a typical case, an unstrapped 8-segment block (e.g. E1198) has four modes in the range 9.1 to 9.5 cm. When strapped, the pi-mode remains at 9.1 cm and the unwanted modes move to 7, 6 and 6.4 cm. Because of their higher frequencies, they will not be excited before the operating voltage has reached that of the pi-mode.]

The introduction of strapping resulted in an improvement in magnetron performance of the same magnitude as that achieved by the introduction of the oxide cathode. For example, E1198, which had previously given around 12 kW peak output with an efficiency of about 10%, gave approximately ten times more output with 200 kW input when strapped.

Even with strapping, there were problems with moding and with variation in efficiency from sample to sample. This was attributed in part to there being minute differences in the dimensions of individual cavities in a multisegment block, and so of their resonant frequencies, an effect which would be increased by idiosyncracies in the geometry of the straps themselves. [In an earlier experiment with an unstrapped magnetron, Sayers had found that 'moding' could be reduced significantly by tuning individual cavities by means of a threaded metal insert.] This problem was analysed by S. Devons of TRE, who in March 1942 introduced 'strap setting' as a method of 'tuning' a multicavity system so that the desired pi-mode of oscillation was enhanced still more at the expense of unwanted modes. Devons also designed and built the first apparatus for this purpose, which involves local distortion of the straps to alter the effective capacitance of each resonant circuit so that the whole assembly has in effect a single very narrow band response in the pi-mode. It also makes it possible to set the frequency of each magnetron to within the specified band, and so improve the yield from a production line.

So far as can be ascertained, 'strap setting' remained secret for the duration of the war, so that British and American magnetrons had a much smaller variation in frequency from sample to sample, and were consistently more efficient than those from other sources.

Continuing research and development of strapped magnetrons at Birmingham University, BTH and GEC resulted in the production of tubes giving up to one megawatt peak in S-band and 200 kW in X-band. It was found that choice of the correct ratio of cathode to anode diameter was also a critical factor in the reduction of 'moding'. This choice could be based only on experiment and accumulated experience.

4.2.4 S-band Developments

From September 1940, when samples of E1189 and E1198 and the first 'Sutton tube' reflex klystrons were available for use in experimental equipment, development of S-band radar proceeded very rapidly. Early in 1941, J.R. Atkinson of TRE and W.E. Willshaw of GEC tested E1198s in a laboratory RDF system at Swanage which detected targets at ranges from 2 to 24 miles. In mid 1941, flight tests were being made on development models of AI MkVII, the first Air Interception radar, which was operational by the end of 1941. As a result of this, several thousand E1198, now type approved as CV38, were manufactured by GEC and BTH in the next two to three years. The large numbers were required partly because of losses in production caused by moding and variations in efficiency from sample to sample, and partly because of unexpected breakage or rejection of good magnetrons in the field, coupled with an over-riding need to keep several hundred equipments flying. [The failures in the field were caused partly by the lack of experience of the operators and the difficulty of 'setting up' outside the laboratory or test room, and partly because the high voltage supply could be switched on by the operator before the cathode had reached its initial operating temperature, so causing flashover before oscillation.]

At the time of the E1198 trials, a working party from Signal School visited Swanage and built an experimental system using E1189. This became the functional prototype of Type 271, a shipborne radar engineered and built by Admiralty workshops and installed in HMS 'Orchis' by the end of March 1941. It gave ranges of 12,000 yards on ships and 1300 yds on a submarine periscope extended to 8 ft. By the end of 1941, fifty Type 271s were operating in escort vessels.

In late 1940, when it was decided to use a magnetron in the Army GL3 gun-laying equipment, the power available from the first generation S-band tubes was insufficient to meet the operational requirement, it being estimated that at least 100 kW peak would be needed.

Thus there were two over-riding priorities in continuing research and development:

(a) to increase efficiency* and power output (to increase range), and
(b) to minimise losses in production and in the field. [Copper and other metals used in the fabrication of magnetrons had to be imported, and could be in very short supply. Apart from that, the available manufacturing facilities were stretched to the limit, and sometimes unable to meet the demand]

---

*In airborne equipment, the magnetron input power is limited by the capacity of the aircraft generating system. Apart from that, the need to operate at high altitudes and with high pulse recurrence frequency to ensure adequate illumination of a target by the scanning radar precluded the use of a mercury thyratron modulator switch. In the early AI systems, the only available tetrode modulator limited the peak power input to the magnetron to about 10 A at 13 kV, but could operate at 2,500 p.p.s. Later on, development of the trigatron switches CV85 and CV125 raised the peak input to 200 kW and then 500 kW, but for adequate operating life the pulse recurrence frequency was limited to about 1500 p.p.s. (see 6.6)

The invention of 'strapping' by Sayers in July 1941 was to provide the means for a major improvement on both counts. Until then, attempts were made to increase power output by development of larger versions of the CV38 (E1198) and NT98 (E1189). In late 1940, an 8-segment magnetron (E1287) with a block 60% greater in anode and cathode diameters and twice as long as E1189 was developed jointly by GEC and BTH for operation with up to one megawatt peak input in an attempt to meet the GL3 requirement. Powers in the range 60 to 100 kW were obtained from laboratory preproduction samples. Some were used in the early development of GL3, including the first operational prototype delivered to ADEE in May 1941, but problems with moding and matching and variations in efficiency from sample to sample made it unacceptable for use long term. In September 1941, a 10-segment tube (E1267) with a larger anode and cathode diameter gave about 150 kW peak 75 W mean with 15% efficiency and much better stability than E1287. Manufactured by BTH and Type Approved as CV41, it was used in the early production models of GL3. [In the course of these developments and preproduction, it was found by BTH that high accuracy of machining and a near perfect internal finish of the block were essential to the achievement of the maximum efficiency and stability.] After CV41, there was no more development of unstrapped S-band magnetrons. In January 1942, BTH obtained 400 kW peak output with 40% efficiency from an echelon strapped (see below) version of CV41. Later Type Approved as CV120, it was used in the production run of GL3.

In late 1941, following Sayers' demonstration of the effect of strapping on E1198, W.E. Willshaw*, A.G. Stainsby and E. Kettlewell of the magnetron development team at GEC completed development of an 8-segment magnetron based on NT98. This gave 80 to 100 kW peak output power with an efficiency around 40% at 15 kV 15 A input. Type Approved as CV56 for use in the Type 271Q Naval radar, it was the first strapped magnetron to go into production. A higher power version, CV76, used the same anode structure with a higher magnetic field. Operating at 25 kV 40 A it was used as the transmitter in Type 277, which was designed by Signal School and used by all three Services in early warning systems from early 1943.

*By this time, Willshaw had taken over from Megaw as leader of the GEC magnetron development team. He succeeded Megaw as leader of the group in 1943. From early 1942, Megaw had concentrated on research projects including studies of propagation in the microwave band and of magnet design.

Early in 1942, Sayers and the group led by L.Rushforth and R.Latham at BTH completed development of BM717, an 8-segment magnetron designed to supersede CV38 in airborne equipment. Operating at 13 kV 10 A [the output from the only available modulator, the tetrode E1155] it gave upwards of 40 kW peak output, about four times that of CV38 under the same conditions. Sayers fitted

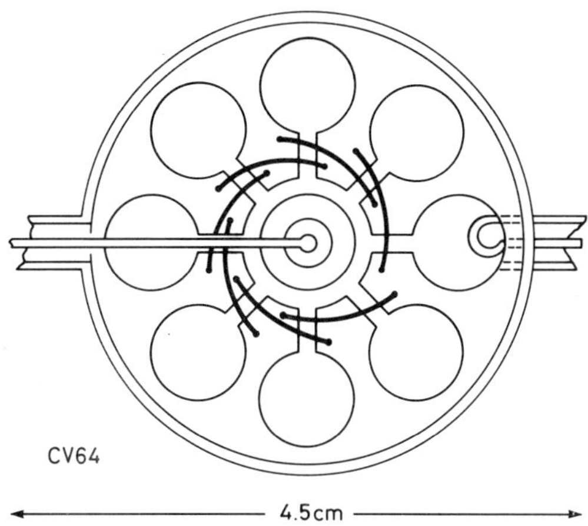

Echelon strapped magnetron

BM717 with a new form of strapping ('echelon'), in which each segment is joined to the next but one by individual heavy gauge copper wire hoops. This resulted in an increase in efficiency and stability compared with the ring strapping used in earlier developments. It also made it easier to fabricate and fit the straps, and to use the strap setting technique developed by Devons. Type Approved as CV64 for use in AI MkVIII, it was the first magnetron to which this technique was applied in large scale production. When the CV85 trigatron switch became available in mid 1942, the peak power input rating was increased to 160 kW. Fitted in the Modulator 64, the uprated CV64 was approved for use in H_2S in October 1942. In late 1943, a higher power version, CV192, was developed to take advantage of the 500 kW peak power available from the trigatron CV125 in Modulator 158.

Early in 1944, CV160, a variant operating with up to one megawatt input, was developed for use in ground radar as an alternative to CV76.

4.2.5 X-band Developments
By mid 1941, when development models of S-band airborne radar were under test, it was concluded that a move to X-band was essential to improvement in performance of AI and ASV, particularly for the Fleet Air Arm. It was planned to fit magnetrons giving at least 20 kW peak output in equipment to be available in 1942.

Attempts were then made by Birmingham, BTH and GEC working together to develop tubes with a performance comparable with that of S-band types, using anode blocks with up to 22 segments to avoid a large increase in magnetic field, and anode bores and cathode diameters comparable with those of 10 cm tubes to maintain the power input rating. These were unsuccessful, partly because the larger number of segments exacerbated moding problems, and partly because of actual and anticipated difficulties of fabrication in numbers. By autumn 1941, when development of the experimental models of the equipment was well advanced and production was planned for 1942, it was decided that as an interim

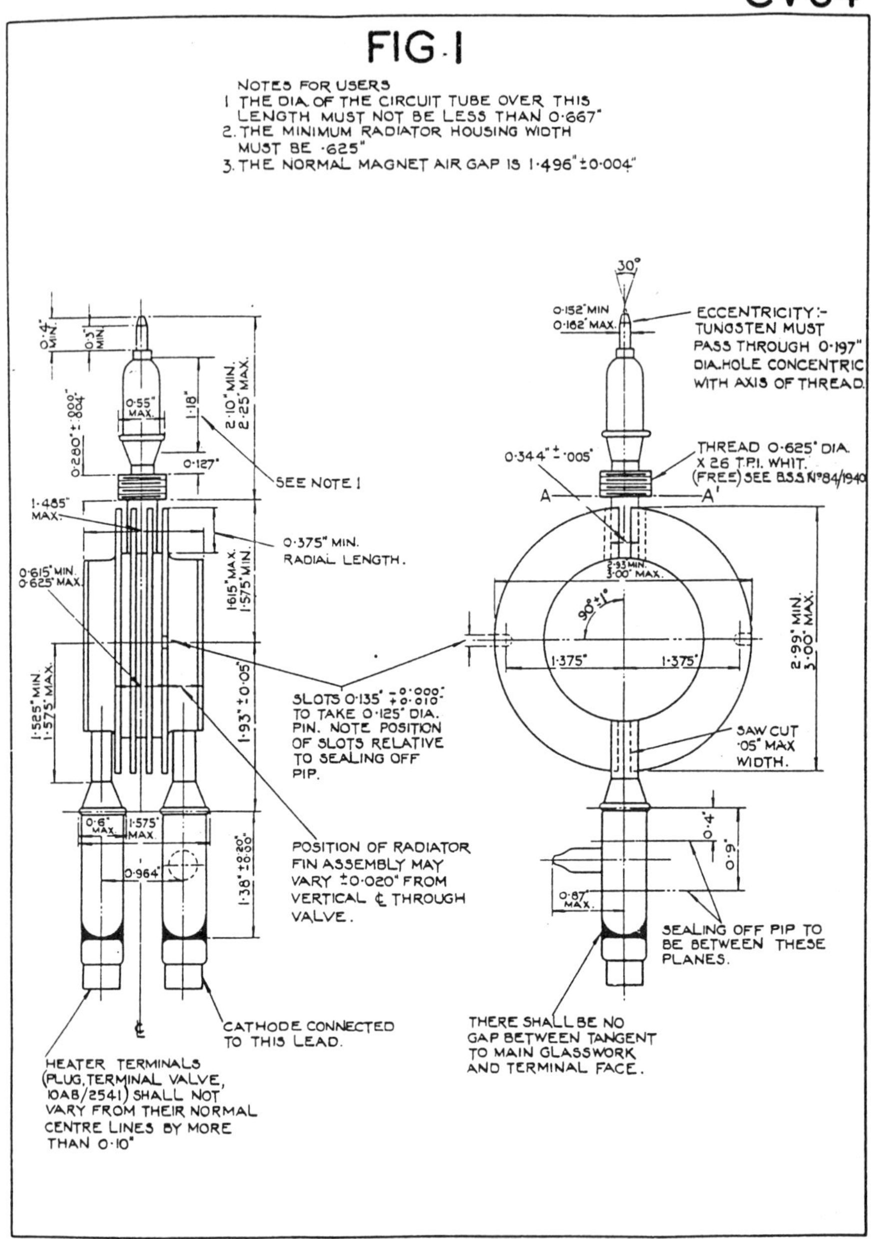

measure, BTH and GEC would make quantities of the best magnetron that had been tested. CV108 had a block with twelve 'slot' resonators, [this configuration was chosen because the slots could be cut accurately by available gear cutting machines], an anode bore of 10.5 mm and a cathode diameter of 5.75 mm (comparable with the dimensions of the S-band CV38), and gave efficiencies up to 20% with 150 kW peak input. The magnetic field was 2670 Oe, about 2.5 times that of equivalent S-band tubes, but available from permanent magnets which were small enough to be fitted in airborne equipment. Although this performance could be obtained with reasonable consistency by careful adjustment in the laboratory, CV108, like CV38, was difficult to set up in the field, and had the same problems in production. As before, large numbers had to be made to keep operational AIX and H_2S equipment flying. Production continued until April 1944.

Following the successful development of CV56 and CV64, attempts were made by Birmingham, BTH and GEC to develop strapped X-band magnetrons with comparable performance, but with little success apart from a few laboratory samples.

At about the same time, the Bell Telephone Laboratories, Westinghouse and Raytheon were attempting to do the same in USA. As part of their programme, BTL had formulated 'Scaling Laws' for magnetron design, so that a tube operating successfully at one wavelength could be replicated at another by reducing the dimensions in proportion to wavelength and increasing the magnetic field in inverse proportion. This presupposed that mechanical considerations would not preclude this process. Using this technique, BTL and Western Electric developed the 725A, a double ring strapped 12-segment X-band magnetron operating at 14 kV 10 A with an efficiency around 35% in a magnetic field of 5000 Oe. Detailed information on this design was available in UK at the end of 1942.

[American manufacturers had also developed new fabrication technology, including the use of 'pull broaches' to form the resonator slots and anode bore in one operation with an accuracy and finish that could not be matched by any other machining process. (This technique was brought to the UK by Rushforth and established first at BTH for production of CV209, and also used by GEC.) In 1943, American magnetron makers developed the technique for cold 'hobbing' and electro-polishing magnetron blocks, in which a tool with the required profile is pressed hydraulically into a short cylinder of pure oxygen free high conductivity copper to form the resonators and anode bore in a single operation, so eliminating the need for additives such as tellurium, which were used to improve the machining properties of copper. This process was also adopted by British magnetron makers. Apart from fabrication processes, US magnetron makers had access to materials technology superior to that available in UK, including those used in the manufacture of copper bar stock for anodes and for production of permanent magnets, so that they were less inhibited by the need for narrow resonator slots and higher fields at shorter wavelengths. Ten years later, a leading British magnetron manufacturer who had contracts from the US Navy for supply of the 4J50 packaged magnetron, could meet the US specification only by importing magnets from USA.]

Stimulated by the BTL success, Sayers (who had visited USA to discuss American magnetron developments) and BTH developed MX57 and MX52, basically identical 12-segment double ring strapped tubes which could be manufactured as replacements for 725A and as plug-in replacements for CV108. [TRE had previously stated that because of the delay in British development, they were prepared to adapt their transmitter to use the 725A, which had a cathode support seal assembly different from that of CV108.] The first samples of MX52 were tested in June 1943. As in 725A, the anode bore (5.6 mm) and cathode diameter (3 mm) were about half those in CV108. Operating at 11 A 13 kV, the efficiency was around 35% with a magnetic field of 5000 Oe. Accepted

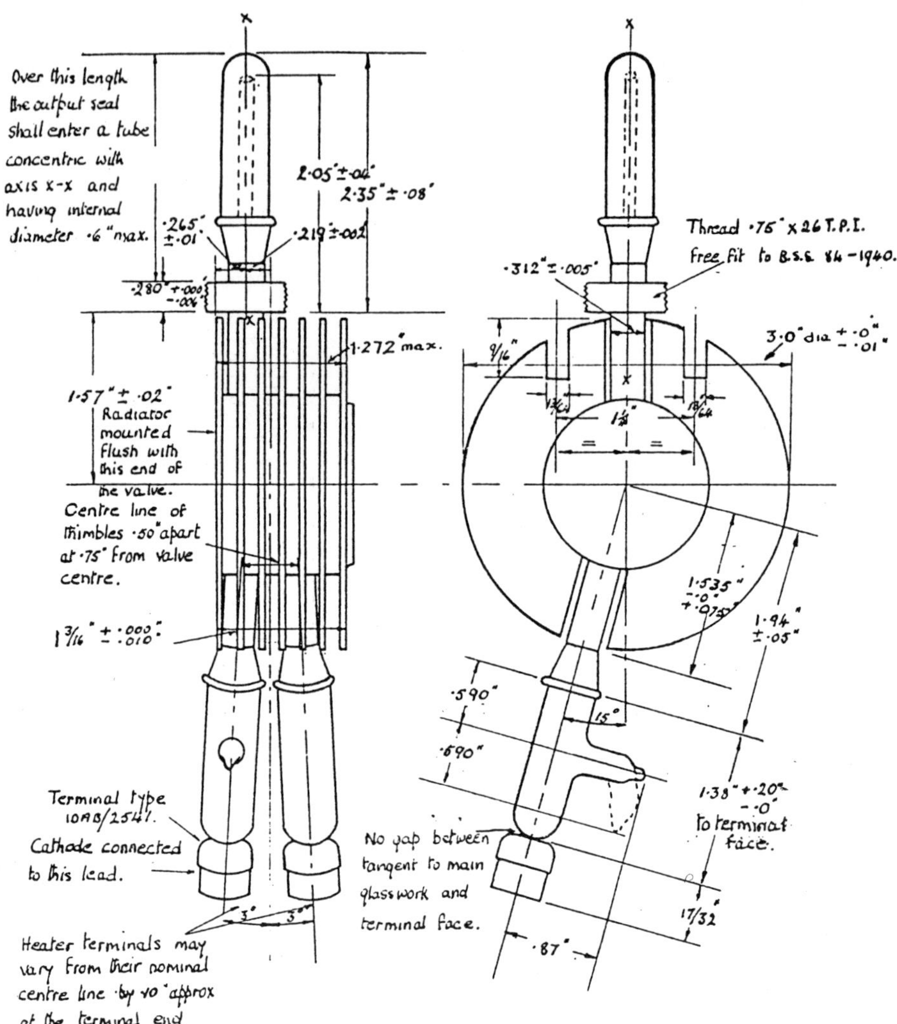

C.V.208.

by TRE in preference to 725A and manufactured as a plug-in replacement for CV108, MX57 was fitted with iron polepieces so that it could operate in the same permanent magnet. Type Approved as CV209, it replaced CV108 in airborne equipment from early 1944.

In August 1943, the GEC team completed development of E1487, a 12-segment double ring strapped magnetron also designed as a plug-in replacement for CV108. The anode bore and cathode diameter, and the anode length, were

CV 209

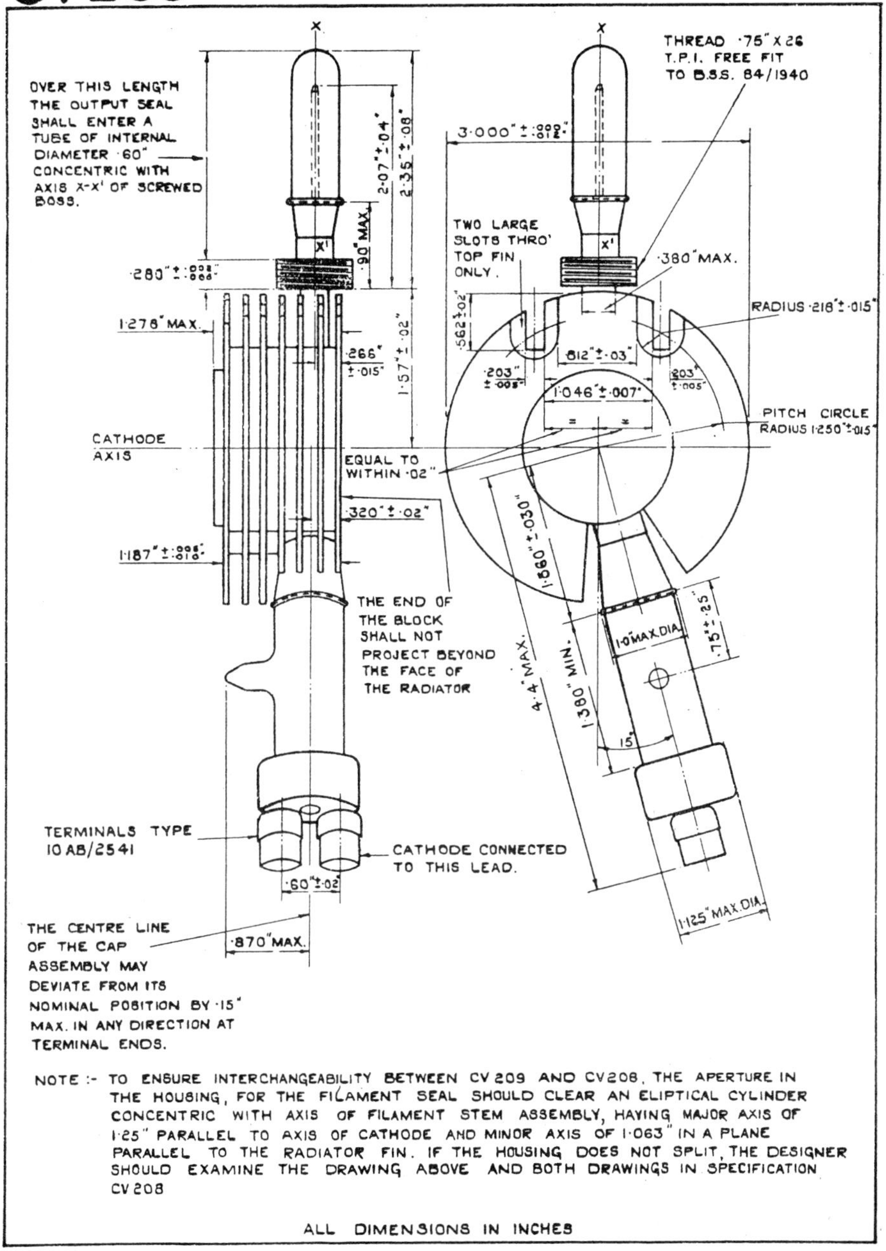

larger than those in CV209, so that its mean power rating was also higher. Operating at 10 A 15 kV the efficiency was around 30% with a magnetic field of 3250 Oe. Accepted by TRE for use in AI and H_2S, it was in preproduction at the end of 1943. Type Approved as CV208, it was fitted with a single iron polepiece so that it could operate in the same magnet as CV108. A later version, CV214, was used in Naval X-band radar. Fitted with an integral waveguide output system, it could be 'set up' during production testing and connected without further adjustment in the field to the equipment waveguide assembly. It had a moulded glass 'boot' similar to that on 725A to protect the heater seals.

4.2.6 K-band Developments

From late 1941, the Clarendon Laboratory at Oxford was engaged in research and development of magnetrons operating at wavelengths below 2 cm, the objective being to improve still further the resolution of airborne radar systems such as H_2S. By July 1942, they had obtained 10 kW peak with about 15% efficiency from a 10-segment unstrapped magnetron operating at 1.6 cm. In February 1943, in line with a decision made by the U.S. Defence Departments, the wavelength was changed to 1.25 cm. In December 1943, powers around 12 kW peak were obtained with 10% efficiency. Development continued, but no British operational radar equipment was designed for the 1.25 cm band because radiation at this wavelength is subject to extremely high selective attenuation by water vapour in the atmosphere. [It has to be assumed that this was not known at the time to those who made the decision to shift from 1.6 cm.]

[The Navy used a Canadian built 1.25 cm system for location of shell splash, where high resolution and limited range could be a tactical advantage.]

The most important outcome of the Clarendon research was the invention of the 'rising sun' magnetron by B.V.Rollin in June 1944. In this design, separation of the pi-mode from the unwanted modes is achieved by having adjacent resonators unequal in length. The mode separation is independent of the length of the anode provided no axial resonances occur, and so it can be longer than a strapped system at any given wavelength. It is also possible to use a larger number of segments (typically up to 18) and so a larger anode bore and cathode diameter.

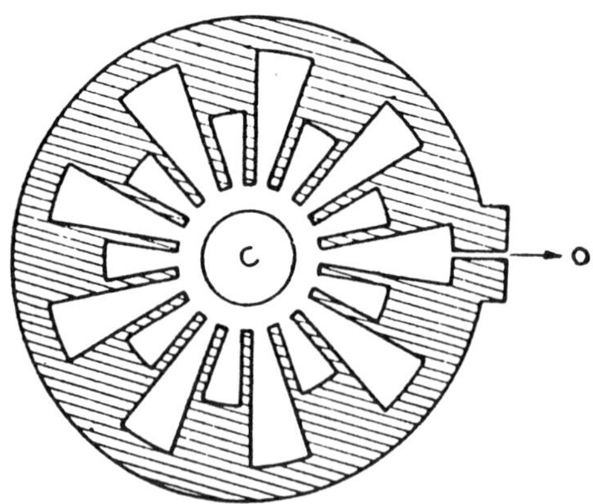

The rising-sun anode.

The power output is coupled through a slot in one long resonator to a short waveguide extending radially outwards and closed by a low loss glass window sealed into the coupling flange at its end. Thus rising sun magnetrons can be designed to give much higher powers than strapped alternatives. This made possible the future development of high power magnetrons at wavelengths below 1 cm, e.g. at Q-band (8.6 mm), which were used in precision radar systems such as H2Q. [In parallel with the work at the Clarendon Laboratory, rising sun magnetrons were developed by Columbia University, New York, who had almost certainly arrived at the design independently. One of their notable achievements was the AX9, a 3 cm tube giving up to one megawatt peak output.

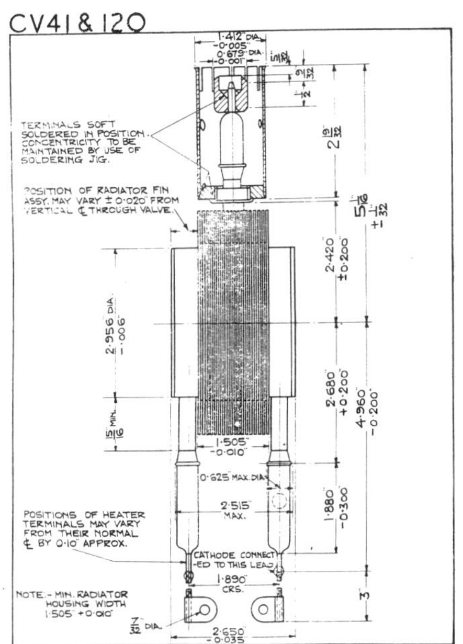

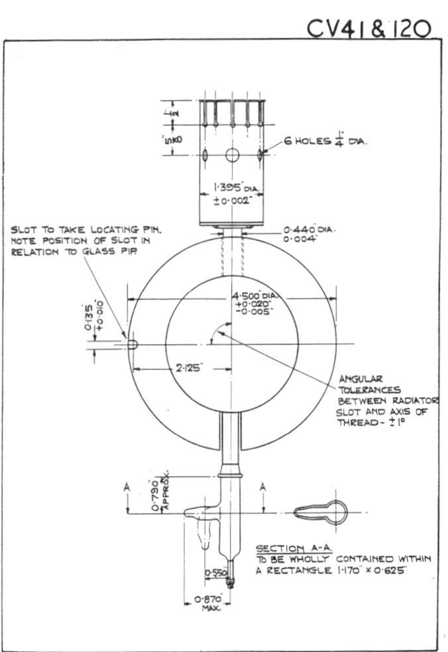

CV 208

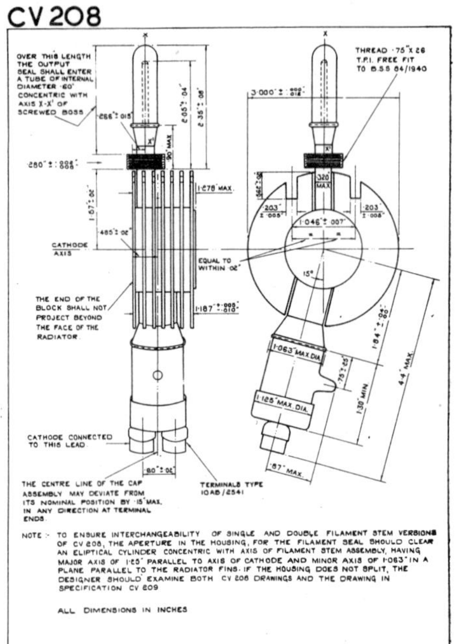

CV 108

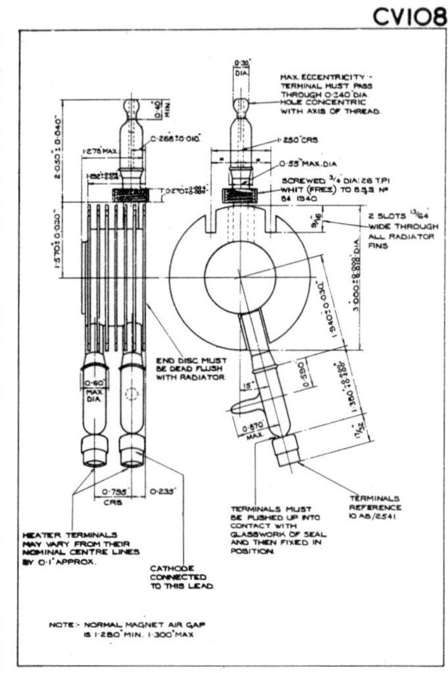

CV 38

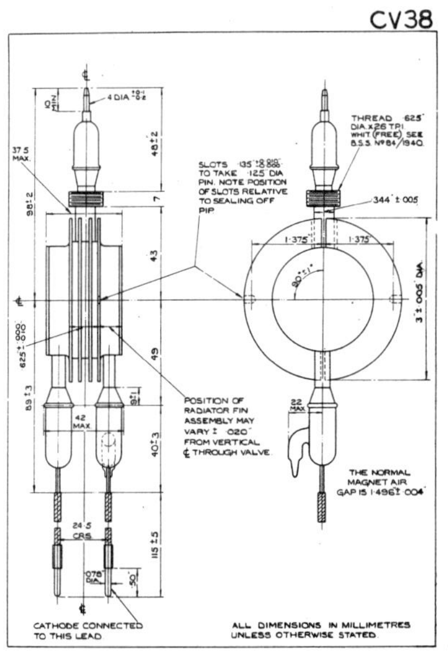

CV 214

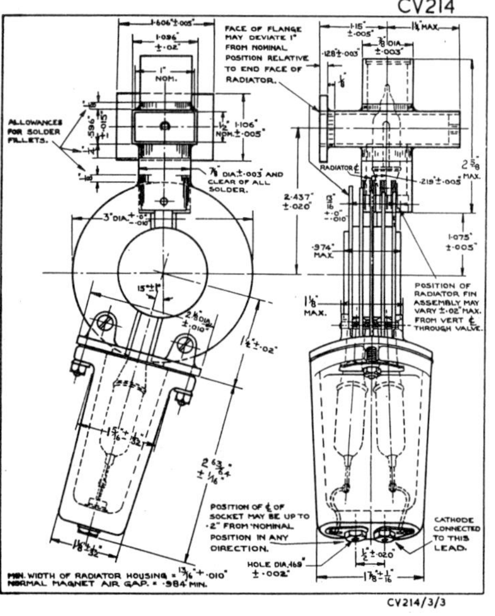

4.3 Reflex Klystrons

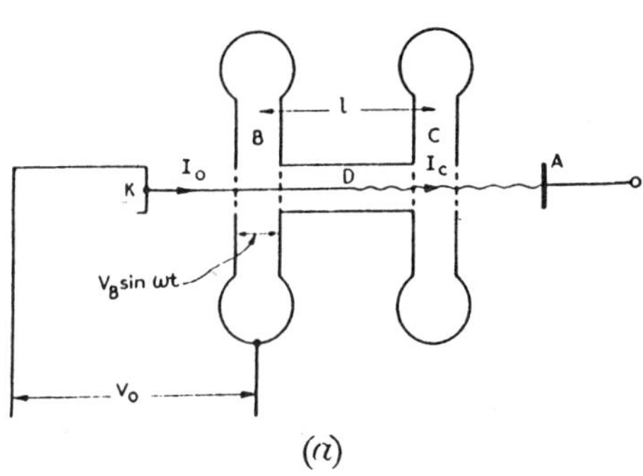

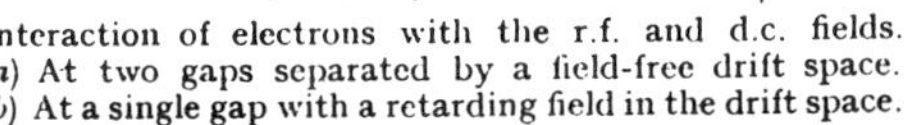

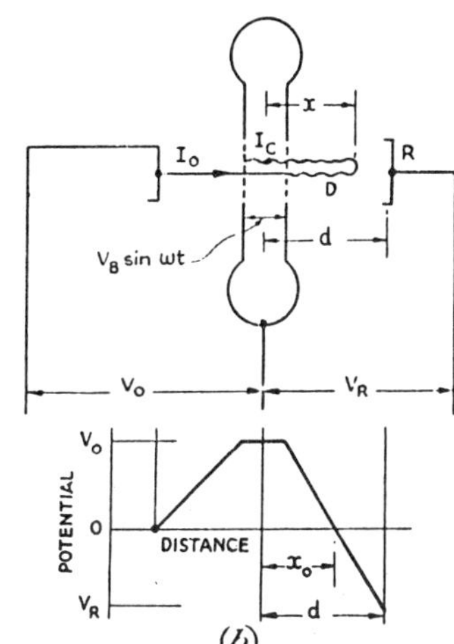

(*a*)

(*b*)

nteraction of electrons with the r.f. and d.c. fields.
a) At two gaps separated by a field-free drift space.
b) At a single gap with a retarding field in the drift space.

4.3.1 Foreword

The first proposals for generating power by means of 'velocity modulation' were made by A.A. and O. Heil in 1935. In 1939, R.H. and S.F. Varian of Leland Stanford University in California applied this principle to the design of 'klystrons' in which an electron beam passes through a pair of 'rhumbatron' resonators*. The first or 'buncher' cavity, to which the RF input signal is applied, causes alternate acceleration and retardation of the electrons as they pass through the aperture into the 'drift space', where the variation in velocity produces 'bunching' of the beam as the slower electrons are overtaken by those which gained in velocity. The bunched beam then passes through the aperture in the second resonator (the 'catcher'), and induces an alternating current therein, so generating an output signal before the beam is collected by a positive electrode beyond the second cavity. The output power is provided by the reduction in velocity of the beam as it traverses the second resonator.

*The rhumbatron resonator was invented by W.W.Hansen in 1939. It resembles a hollow 'ring doughnut' with a horizontal circular slot in the wall of its central aperture. In geometric terms, an ideal rhumbatron is the surface of revolution of a 'Hertzian dipole' resonator, which is a circular wire loop (the inductance) with a gap between a pair of small discs or spheres attached to the ends of the wire (the capacitance). By using the toroidal form, radiation losses are virtually eliminated, and so the circuit has the high 'Q' and high impedance required for efficient conversion of the energy in the beam into RF power, which can be taken from the toroid by a coupling loop or through an aperture in the outer wall. In its practical form, the rhumbatron is flattened, taking the form of a short hollow cylinder with a relatively thick outer wall to which the output coupling can be attached, and a central aperture defined by the remaining part of the toroidal surface. Electrons pass through the central aperture in a small fraction of the period of oscillation of the RF signal and interact with the electric field fringing the gap. The coupling to the beam can be increased by attaching grids with high transparency to each 'horn' of the resonator.

4.3.2 The Reflex Klystron

If a klystron is to operate as a tunable oscillator, it is obviously an advantage to use a single resonator and arrange that the beam make a second transit in the correct phase by means of a reflector electrode which determines the time spent by the beam in its double traverse of the drift space. This is the principle of operation of the 'reflex klystron' which was described by the Stanford group in 1939.

[In the Heil tube, the single resonator is a tuneable coaxial line with a strip electron beam which makes two transits of the gap between the outer conductor and the inner, which contains the drift space. Because of its higher operating current and the lower 'Q' of its resonant circuit, the output signal from a Heil tube has a higher noise content than that of a reflex klystron, and so it was not used as a local oscillator in microwave receivers. Developed by STC, it was used extensively in communications and measurement apparatus.]

Shortly after the decision to develop 10 cm radar was made in early 1940, R.W. Sutton and the Signal School group at Bristol University started development of a reflex klystron based on published information from USA.

[The power input from the local oscillator to the mixer in a microwave receiver is many times that of incoming signals. Thus the noise factor of the system will be dependent upon the noise content of the local oscillator output. This is determined partly by the shot noise in the beam and partly by the selectivity of its associated resonant circuit. There is virtually no space charge smoothing of the beam current, and so the local oscillator must operate with a power input no greater than that required to give a power output sufficient to minimise frequency conversion loss in the mixer. Because the noise content of the output from the oscillator is proportional to the band width of its output circuit, this should have high 'Q', which is a measure of its selectivity. Thus the combination of a lightly loaded rhumbatron resonator with a high impedance electron beam offered the best prospect for successful design.]

In Robert Sutton's own words:-

'Although it seemed rather a gamble at the time, the reflex klystron, on account of its high Q cavity, appeared to us to be the best bet for a local oscillator. The cavity could in theory be tuned by distorting it in some way, but to distort a cavity in a reproducible manner inside a vacuum seemed hardly practicable.

At this stage we saw being developed by GEC at Wembley a new type of glass-to-metal seal. It consisted of a copper disc sealed between two glass tubes. This immediately showed a way of having a cavity resonator with part of it outside the vacuum and thus accessible to some form of tuning. GEC had trouble with the thickness of the disc; if it was too thick the glass cracked, and if it was too thin the copper buckled. So there was an optimum thickness at which the glass did not crack too easily. We got over this difficulty by using thin copper and stiffening it at the point of seal by pressing a deep annular corrugation into it. This made the seal perfectly practical.

The oscillator was made using two of these disc seals with a metal ring, in two halves, clamped between the discs. An electron gun, using a standard cathode ray tube cathode assembly and two cylindrical focussing electrodes, was sealed into one end of the tube and a hollow cylindrical reflector electrode into the other.

One small point which we discovered the hard way was that the contact between the copper discs and the ring had to be uniformly good all the way round; any area of bad contact drastically reduced the Q. The clamping had to be tight enough to deform the copper all the way round.

Tuning was first attempted by screwing large plugs in and out of the resonator ring. Even when silver plated they were far too noisy to be of any use. They were retained purely for presetting the range, but they had to be fitted with lock nuts screwed down tight. It was obvious that some kind of non-contact tuner had to be devised. The tuner consisted of a thin walled tube as wide as the resonator, fixed to it radially outwards. A metal cylinder moved in and out of this tube for tuning. It had the smallest clearance that it could safely be given. It was moved by a small diameter threaded rod in a fixed nut. Luckily

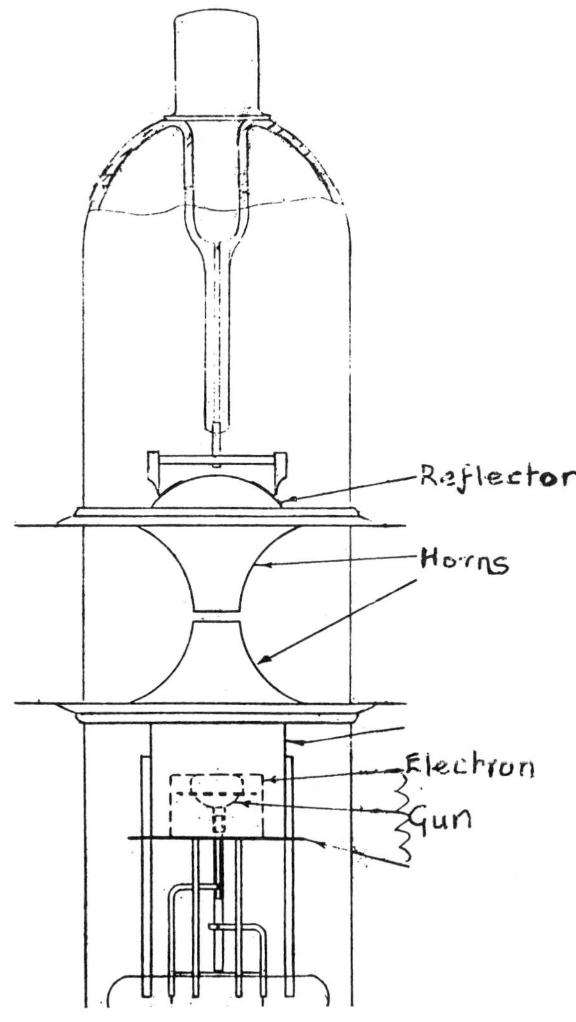

'Sutton tube'

the contact at this much smaller diameter was perfectly quiet. Actually the tuning plug must have been about a quarter wavelength, so we had, without realising it at the time, an effective filter between the resonator and the screw thread.

These oscillators had the disadvantage that the gun voltage was critical and it had to be varied with the tuning. This meant coupling a potentiometer to the tuning knob. Later on, EMI made a great improvement in the valve by using a much smaller reflecting electrode and putting it very close to the resonator. These valves covered the whole tuning range with no change in voltage being needed. They also gave a much larger output. The reduced drift length was clearly more than offset by the greater fraction of the electron beam getting back through the resonator.'

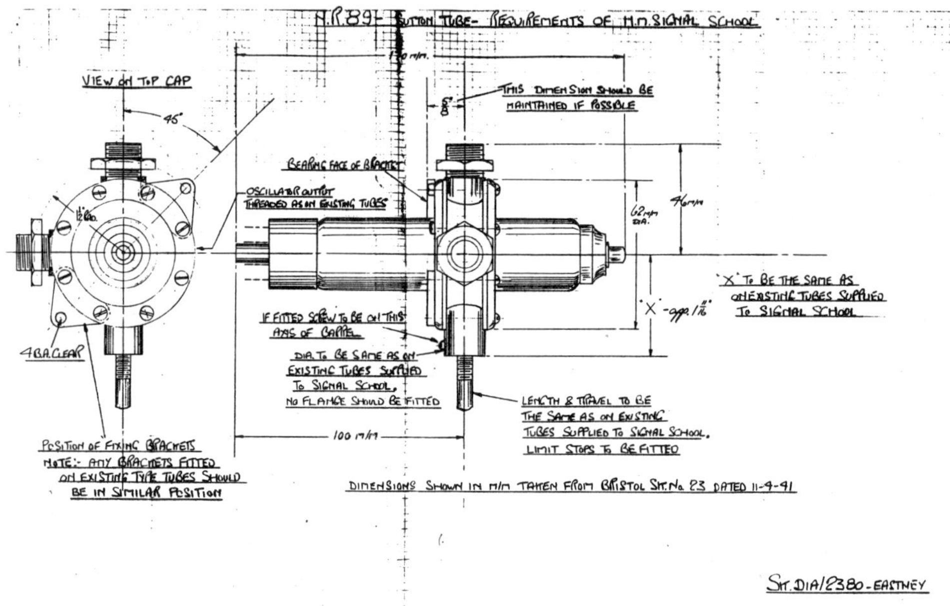

Designed by A.M. Reith and J. Thomson, the first 'Sutton tube' operated successfully in September 1940, giving tens of milliwatts output at 10 cm with a 3% tuning range. Development was completed in December 1940. Known as NR89 with an 8% tuning range, the 'Sutton tube' was tested by H.W.B. Skinner in his experimental radar at Swanage in early 1941. In small scale production from March 1941, it was used in the early Type 271 Naval radar, being manufactured in quantity to the Signal School design by E.M.I. Ltd and E.K. Cole. Frequency variants were made for the Army GL3 equipment and for airborne applications.

4.3.3 S-band Developments

Although the NR89 made possible the development of sensitive 10 cm radar receivers, it was difficult to use because of the high operating voltage (about 1700 V) and its sensitivity to variations in the anode and reflector voltages and in that applied to the focussing electrode in the electron gun. This arose in part from the choice of a comparatively long drift time for the electrons in the reflector space to achieve the degree of bunching thought necessary for efficient operation.

As a result of theoretical studies by N.C. Barford and M.B. Manifold and experimental work by A.F. Pearce and B.J. Mayo of L.F. Broadway's group at EMI, major changes were made to the design of NR89. The diameter of the interaction gap was reduced from 5 mm to 2 mm, which increased the coupling between the electron beam and the resonator, and the drift time of the electrons reduced from $6\frac{3}{4}$ to $1\frac{3}{4}$ periods of oscillation by changing the shape of the reflector and reducing its separation from the exit aperture in the rhumbatron. The electron gun was also redesigned and simplified so that the focussing electrode operated at cathode potential. Experimental work was completed in June 1941, and samples of the new design (later known as CV36) were delivered in November to the team developing GL3 at BTH. Like NR89, which they replaced, CV36 and its frequency variants CV35 and CV67 were high voltage tubes. Operating at 1400 V with the reflector voltage between -230 and -320 V, they gave 300 mW output with 10 W input. The normal tuning range was about 5%, but the three tubes between them would cover the band 8.9 to 10.9 cm. Three tuning plungers were used, two for pre-setting the wavelength and the third for tuning over a 2% band. Unlike NR89, the new S-band tubes were relatively insensitive to variations in the applied voltages.

Although tubes in the CV35 series were easier to use than the NR89, they still had the disadvantage of operating at high voltage, particularly in airborne equipment. There were also difficulties in maintaining the required difference in frequency between the magnetron and the reflex klystron as they were subjected to fluctuating ambient temperature during warm up in the first few minutes of operation, or during changes of operational altitude. These problems led to a request by TRE in late 1941 for the development of a tube to operate below 300 V and provide at least 20 MHz variation in output frequency by adjustment of the reflector voltage. [Variation of the reflector voltage about the optimum also produces a reduction in power output, and so the 'electronic tuning range' is defined as the difference of the frequencies at which the output power falls to one half of the maximum.]

Based upon theory which had been developed by EMI before and during the development of CV35, the new tube was designed to use a similar form of construction with two major differences in design. The shape of the reflector was modified to provide a longer drift space and so increase the electronic tuning range. The apertures in the horns of the resonator were increased in diameter by about two times so that a much higher beam current could be used, and fitted with highly transparent grids to maintain the coupling between the beam and the resonator. Development of KR6 started in mid 1942, with assistance from the Signal School group at Bristol, who built some of the first experimental tubes and devised a technique for making the grids by folding and brazing a very thin

CV 67

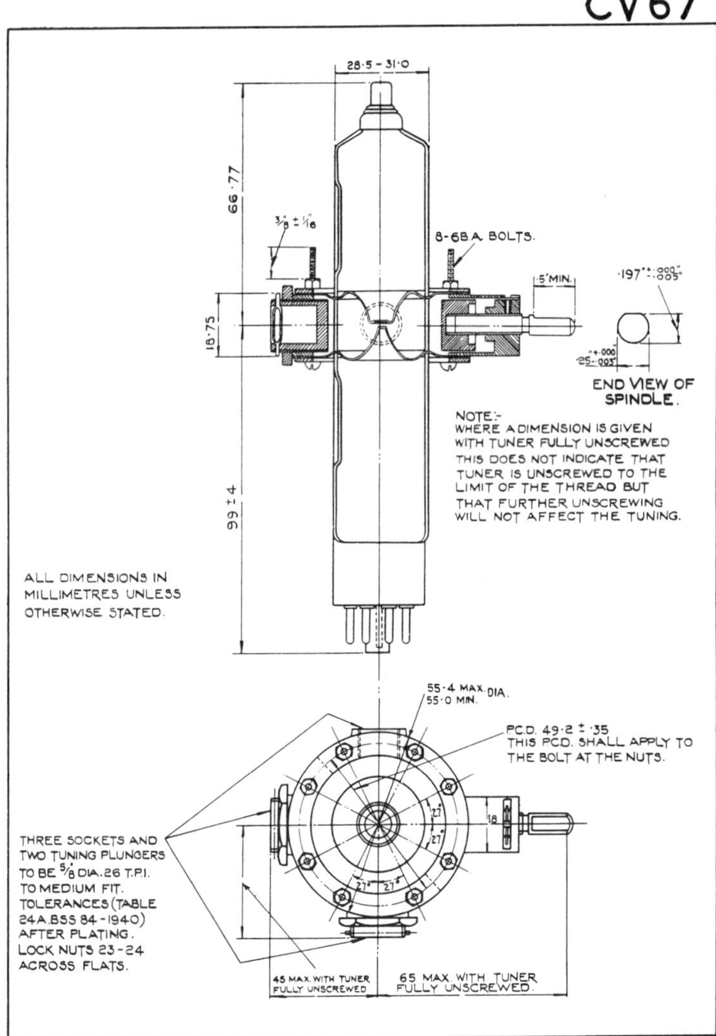

corrugated tape about 1 mm wide. Because it operated at about five or six times the current of CV35, and had grids in the path of the beam, there was some concern that KR6 would generate unacceptably high noise (the 'Q' of the resonator was also lower than that of CV35), but by May 1943 it had been demonstrated that the performance of KR6 as a local oscillator was completely acceptable. Operating with 8 W input at 250 V, the power output was about 150 mW with an electronic tuning range of 30 MHz produced by a variation of 25 V about the −150 V applied to the reflector. Type approved early in 1944, CV116 and its frequency variants CV237, CV238 and CV272 covered between them the range 8.3 to 11 cm, and replaced the CV35 series in British S-band radar.

CV116

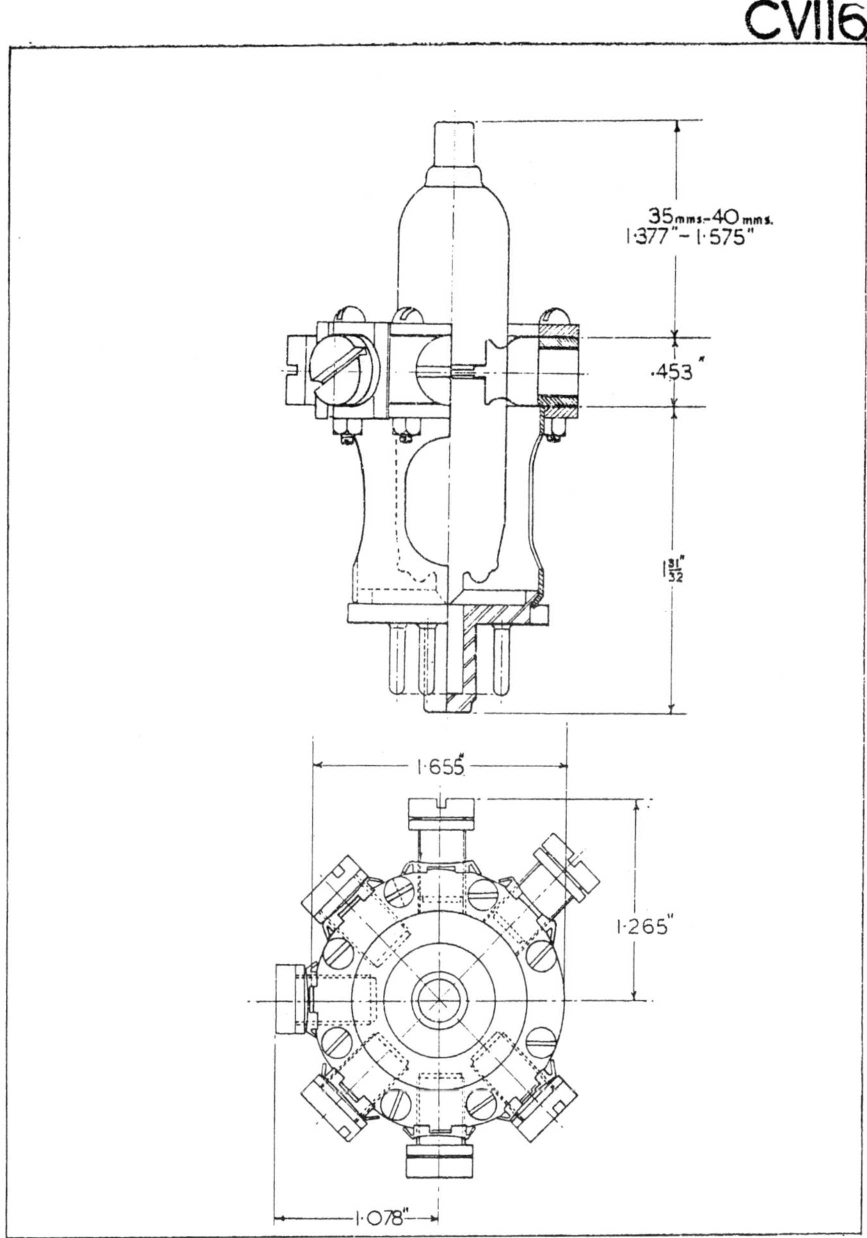

4.3.4 X-band Developments

In late 1940, following a visit to the Signal School group, B. Bleaney, D. Roaf and J.G. Daunt of the Clarendon Laboratory started development of an X-band reflex klystron based upon existing S-band technology. Although the electron gun and reflector assemblies could be scaled down, it was clearly impossible to use a resonant cavity of the same type, partly because of its small size, and partly because the glass would introduce unacceptably high dielectric loss, even if it were possible to fabricate such an assembly. It was therefore decided to use the NR89 assembly techniques in combination with an harmonic resonator* of the type proposed by A.D. Blumlein of EMI. [The Clarendon group was closely in touch

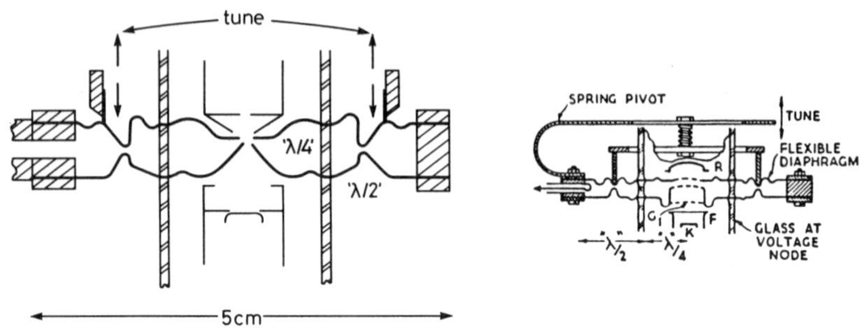

Harmonic resonators

with the work at EMI, and collaborated with them in research and development of reflex klystrons.] Samples were available for experimental use in March 1941, and followed during the next year by the construction of about 100 tubes, many of which were supplied to TRE for use in their X-band developments. When EMI became the main suppliers of X-band types, the Clarendon group concentrated on the research and development of reflex klystrons for operation down to 1.25 cm.

> * In harmonic resonator klystrons, the cavity operates in a 'threequarter wave' mode, with a nodal circle surrounding the interaction gap, and an antinodal circle between this and the outer wall. The glass work is sealed to the discs at the nodal circle and so does not introduce unacceptable RF loss. The resonator is tuned by moving one disc up and down at the antinodal circle, each disc being pressed with a 'V' profile which localises the 'capacitance' of the outer ('half wave') element of the cavity, from which the power output is taken via a coupling loop and a coaxial line or through an aperture in the outer wall direct to the waveguide assembly in the equipment.

In parallel with the later stages of the Clarendon activity, for which they supplied tuning mechanisms and electron guns, EMI developed the harmonic resonator design as KRN2, which operated with a slightly lower power input than the earlier tubes so that it could use existing power supplies in airborne equipment. The first preproduction samples were available in September 1942. Operating with 10 W input at 1500 V the power output was about 100 mW, tunable over the range 3.05 to 3.45 cm. Type approved as CV87, it was used in the early X-band AI and H_2S equipments. Because its differential screw tuning mechanism was difficult to manufacture, CV87 was replaced early in 1943 by

CV129, which used an identical active unit mounted in a smaller and simpler tuning mechanism. For mechanical reasons it was not possible to achieve a drift time less than $3\frac{3}{4}$ periods of oscillation, and so CV87 and CV129 were more susceptible to voltage changes than the CV35 series.

Development of CV129 was followed by KRN3, a modification providing electronic tuning. Based upon theoretical work by Bleaney and Barford, the reflector was changed from a flat disc to a hollow cylinder. This reduced the power output, but not unacceptably, and provided an electronic tuning range of 30 MHz. Type approved as CV130 in 1943, it was fitted with a waveguide coupling for direct attachment to the equipment waveguide.

CV87(KRN2)

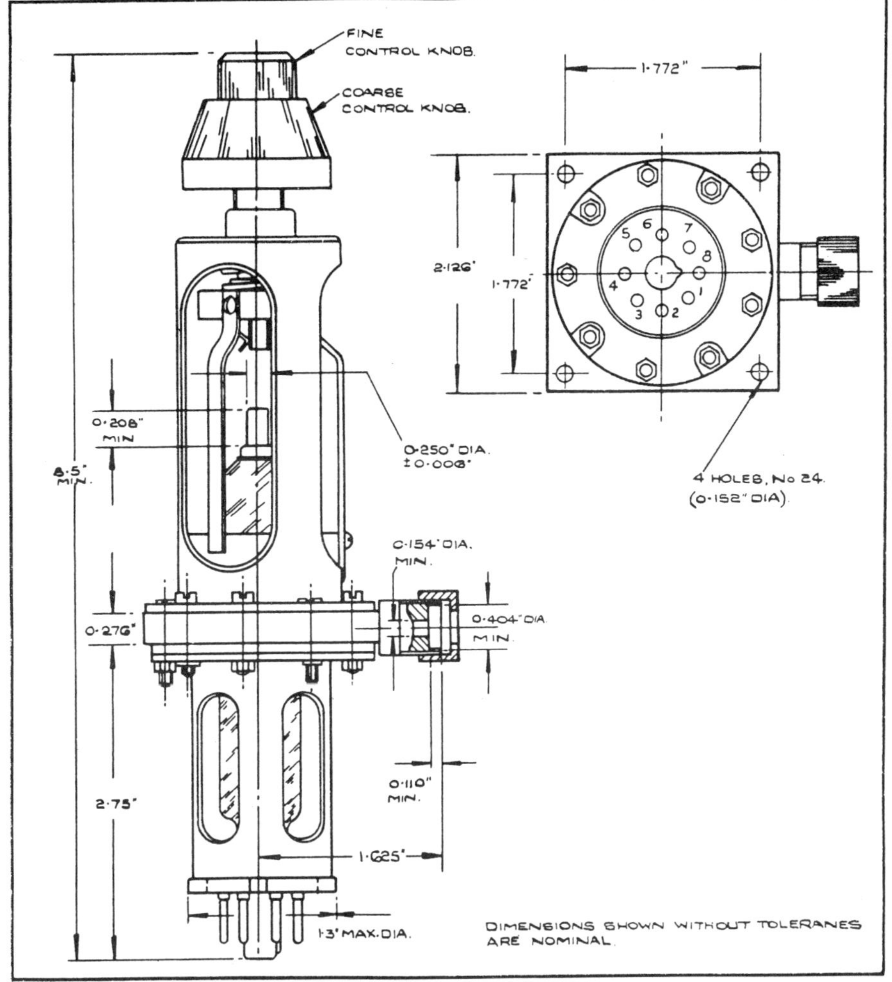

CV 129

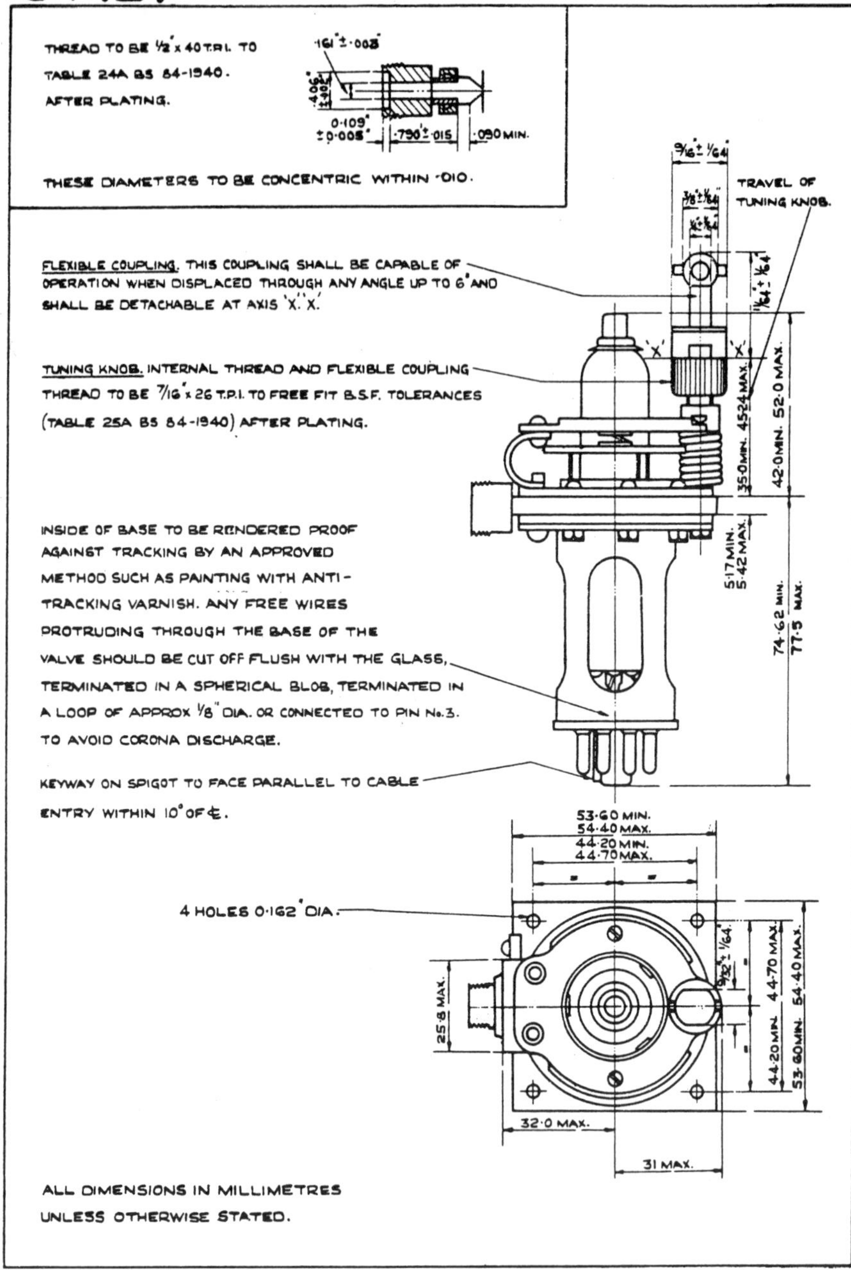

THREAD TO BE 1/2" x 40 T.P.I. TO
TABLE 24A BS 84-1940.
AFTER PLATING.
161 ± ·005
·406 ± ·005
0·109 ± 0·005"
·790 ± ·015
·090 MIN.
THESE DIAMETERS TO BE CONCENTRIC WITHIN ·010.
9/16 ± 1/64
TRAVEL OF TUNING KNOB.
1/8 ± 1/64
1/64 ± 1/64
FLEXIBLE COUPLING. THIS COUPLING SHALL BE CAPABLE OF
OPERATION WHEN DISPLACED THROUGH ANY ANGLE UP TO 6° AND
SHALL BE DETACHABLE AT AXIS 'X.' X.'
TUNING KNOB. INTERNAL THREAD AND FLEXIBLE COUPLING
THREAD TO BE 7/16" x 26 T.P.I. TO FREE FIT B.S.F. TOLERANCES
(TABLE 25A BS 84-1940) AFTER PLATING.
X
X
45·24 MAX.
52·0 MAX.
35·0 MIN.
42·0 MIN.
INSIDE OF BASE TO BE RENDERED PROOF
AGAINST TRACKING BY AN APPROVED
METHOD SUCH AS PAINTING WITH ANTI-
TRACKING VARNISH. ANY FREE WIRES
PROTRUDING THROUGH THE BASE OF THE
VALVE SHOULD BE CUT OFF FLUSH WITH THE GLASS,
TERMINATED IN A SPHERICAL BLOB, TERMINATED IN
A LOOP OF APPROX 1/8" DIA. OR CONNECTED TO PIN No.3.
TO AVOID CORONA DISCHARGE.
5·17 MIN.
5·42 MAX.
74·62 MIN.
77·5 MAX.
KEYWAY ON SPIGOT TO FACE PARALLEL TO CABLE
ENTRY WITHIN 10° OF ₵.
53·60 MIN.
54·40 MAX.
44·20 MIN.
44·70 MAX.
4 HOLES 0·162" DIA.
9/32 ± 1/64
44·70 MAX.
54·40 MAX.
25·8 MAX.
44·20 MIN.
53·60 MIN.
32·0 MAX.
31 MAX.
ALL DIMENSIONS IN MILLIMETRES
UNLESS OTHERWISE STATED.

British development of X-band reflex klystrons was overtaken at the end of 1942 by American designs, when the US Defence Departments adopted the Western Electric 723A. This was the first low voltage X-band reflex klystron to be produced in quantity. Operating with about 6 W input at 300 V, it gave about 40 mW output in the band 2.9 to 3.4 cm with an electronic tuning range of

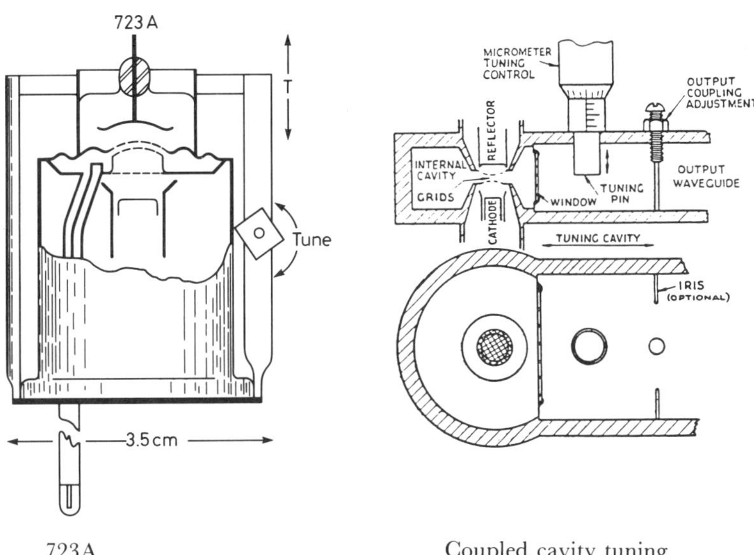

723A   Coupled cavity tuning

50 MHz. Supplies were available for use in British equipment from early 1943. Thereafter, the effort at EMI and BTH, who had been working on all metal low voltage X-band klystrons for about two years previously, was concentrated on developing a plug-in replacement for the American tube. Although they were successful in developing functional replacements for 723A, neither EMI with CV322 or BTH with CSS2 produced a complete plug in equivalent. Within a few years, CV322, CSS2 and 723A were replaced by a variety of British and American all metal tubes with fundamental cavities coupled through a sealed glass window to a short length of output waveguide which contained the tuning plunger and a matching stub, and had at its outer end a standard waveguide coupling flange.

Designed by BTL, 723A was a compact 'all metal' tube with an internal resonator operating in the fundamental mode. The output was taken via a coupling loop and a coaxial line which passed through the base of the tube and then through a vacant hole in the standard 'US Octal' socket into which the tube was plugged, into the waveguide below. There the extended inner conductor of the coaxial line acted as a radiator, protected by an extension of polythene insulation. The resonator was tuned by moving the reflector and the top of the cavity up and down together to vary the spacing of the grids in its aperture, the top of the resonator being part of the vacuum envelope and corrugated annularly. The tuning mechanism was much simpler and cheaper to manufacture than that of CV129. It comprised a rigid pressing extending across the top of the tube from a fulcrum pillar attached to the base of the tube to a diamond shaped strut with its foot also attached to the bottom of the tube. The centre of the pressing, which housed the reflector assembly, was welded to the top of the tube. Tuning was by means of a screw which drew the two halves of the diamond together to increase the resonator gap (the struts had a nut midway up each half). Because of the geometry, the ratio between contraction of the strut and its vertical elongation was about 10 to 1, so that fine tuning was no problem. Because the strut was fabricated from two strips with no pivots there was virtually no backlash. [variants of this system were used in CV322 and CSS2 and in tunable X-band TR cells developed by GEC].

4.3.5 K-band Developments

In late 1941, as part of their programme of development of tubes for operation at shorter and shorter wavelengths, the Clarendon group built and tested successfully a small number of reflex klystrons operating at 2 cm. [TRE had expressed some interest in this band for airborne applications, but shortly after opted to work at 1.6 cm].

By August 1942, up to 15 mW output with about 10 W input at 2 kV had been obtained from a tunable reflex klystron covering the band 1.49 to 1.96 cm. This used a structure similar to that of the EMI KRN2 X-band tube, except that the resonator operated in the fundamental mode and was coupled through a gap in its outer wall to a surrounding secondary resonator in which the glass was sealed to the copper discs at the nodal ring, and from which the output was taken by a coupling loop. A simplified version, designed for making in quantity, used the electron gun and tuning mechanism of CV129, which were supplied by EMI. Samples were tested successfully in February 1943, but in line with the decision of the US Defence Departments to operate at 1.25 cm, work was suspended in April. In June 1943, a number of tubes operating at 1.25 cm had been tested successfully. These had an assembly and resonator structure similar to that of the 1.6 cm tubes except that the inner fundamental resonator was coupled to the surrounding secondary cavity by a low height annular waveguide one quarter wavelength long. In August 1943, the reflector assembly was modified to use a smaller glass tube sealed to a flared copper cylinder which was gold brazed to the upper disc by eddy current heating. Samples and part assemblies of this tube were taken to the US by Bleaney in September 1943, where they were assembled and tested by Raytheon, who later adopted the salient features of its design. A later version (VX302) used a pair of copper plated nickel iron tubes to carry the gun and reflector seals.

By November 1943, the Clarendon group had successfully tested the first samples of an 'all metal' reflex klystron operating at 1.25 cm. In this arrangement, the top and bottom discs of the resonator assembly were part of the vacuum envelope, and formed from copper plated annularly corrugated nickel iron discs, which made it possible to vary the frequency by adjustment of the resonator gap. As in the earlier design, the inner fundamental resonator was coupled to the outer secondary cavity by a low height annular waveguide section one quarter wavelength long. The output was taken from the outer resonator by means of a waveguide comprising a low loss ceramic plug hard soldered into a mild steel sleeve which was inset into the outer wall of the secondary cavity. The gun and reflector assemblies were attached by butt seals to the nickel iron discs. The performance of this tube was similar to that of the earlier design.

Like the 1.25 cm magnetrons, the Clarendon reflex klystrons were not used in British radar equipment because the absorption of K-band radiation by water vapour was well known soon after the successful development of VX302, which gave up to 20 mW output with 50 MHz electronic tuning when operating with 10 mA input at 1800 V.

4.4 Microwave Crystals

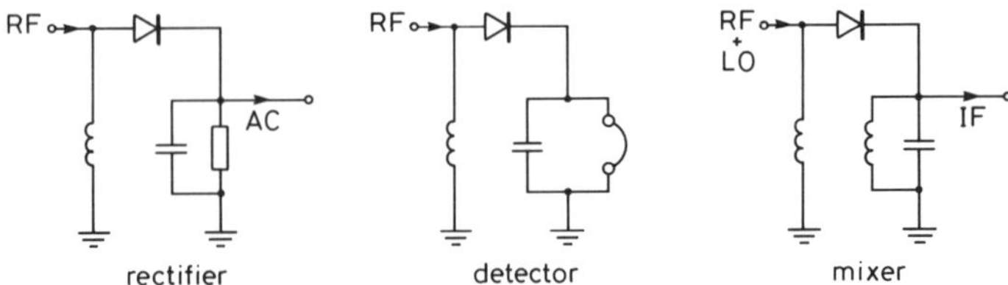

4.4.1 Diode Rectifiers and Mixers

When an alternating voltage is applied to a diode, it conducts during the positive half-cycles so that a capacitor in its cathode lead will be charged to a voltage close to the peak value of the AC input. If a resistor is connected in parallel with the capacitor, the diode output voltage will reproduce a periodic variation in the amplitude of the AC input, provided the time taken for the resistor to discharge the capacitor is short compared with the period of variation. The diode rectifier can then act as the 'detector' of an amplitude modulated input signal. [If the input signal is pulsed, the capacitor discharge time must be a small fraction of the pulse duration.]

> To rectify efficiently, thermionic diodes require an input of at least one volt AC. A point contact rectifier (such as the 'cats whisker' crystal used in the 1920s as the first and sometimes the only stage in domestic radio receivers), will operate effectively with inputs of a small fraction of a volt. No known rectifiers will operate below this level because of the intrinsic potential difference at the interface between the whisker and the crystal. Even if this were offset by a DC bias voltage, there would be virtually no difference between the 'forward' current and the 'leakage' current in the reverse direction.

If the input to the rectifier comprises two alternating voltages of different frequency, that applied to it at any given instant will be their algebraic sum, varying in amplitude at their difference or 'beat' frequency. As before, this variation will be reproduced by the rectifier output voltage, provided the capacitor discharge time is chosen correctly. By replacing the resistor by an inductor to form a circuit resonant at a chosen fixed frequency, and varying the frequency of one of the input voltages, the rectifier and its output circuit will operate as a selective 'mixer' or frequency converter for very small input voltages, provided the amplitude of the swept frequency voltage is large enough for the rectifier to operate efficiently. This is the basic principle of operation of 'superheterodyne' receivers, in which the beat or 'intermediate frequency (IF) is amplified by a chain of fixed-tuned band pass amplifiers before being rectified to obtain the audio or video data carried by the original incoming signal. Because the mixer conducts during alternate half- cycles, there is an intrinsic 'conversion loss' which degrades the sensitivity and noise factor of microwave receivers. [The latter effect is much less in systems operating below 1 GHz, where the first stage in the receiver is a signal amplifier designed to give high gain and low noise factor]

Because of the effect of electron transit time on their efficiency of rectification, thermionic diodes cannot be used as detectors or mixers in microwave receivers.

[As the frequency of the input signal is increased above a few tens of megahertz, the polarity of the input voltage will begin to reverse before electrons leaving the cathode can reach the anode, so causing them to move away from it.]

Even if a diode could be made with an anode-cathode spacing small enough to make the transit time effect insignificant, its cathode area (and so its power input and anode current) would be many times that of the contact area of a whisker on a crystal. Consequently the shot noise in the hypothetical diode would cause a much greater degradation of receiver noise factor than that arising from thermal noise in a point contact rectifier. Apart from that, its anode-cathode capacitance would provide a low impedance parallel path for the RF input power, so reducing the diode rectification efficiency (see 3.4.6).

Thus use of a point contact crystal mixer with a reflex klystron local oscillator offers the only feasible solution to the problem of designing microwave receivers with high sensitivity and an acceptable noise factor.

4.4.2 Early Days

Notes on the History of the Crystal Capsules
H.W.B. Skinner 28th July 1942

1. The idea of reverting to a crystal instead of using a valve mixer for cm waves was due to Professor Oliphant early in 1940. He put effort at Birmingham on improving the crude crystal. Other substances than silicon were tried without very much success. The exterior design of the modern British crystal dates from this time.

2. The first crystals to be uniformly successful as microwave mixers were made by myself at TRE. The essentially new features were:-

(a) Discovery of the DC resistance test, which, though not an entirely accurate test of mixer performance, forms an essential part of the manufacturing process of finding good spots.

(b) The 'tapping process' whereby an initial random contact can be turned into a good spot. The tapped contact is also considerably more stable than the untapped contact.

(c) Wax-filling of the capsules for protection against vibration. (The idea for this may perhaps have come from Professor Oliphant's Laboratory).

3. The TRE capsules were glass envelopes with wires attached to the ends. They were for a time the only successful crystals. For example, TRE was called upon to supply crystals for the earliest 271 Naval Sets early in 1941.

4. The glass crystals (capsules) were not considered suitable for commercial production and also the capsule of the type suggested by Professor Oliphant to BTH was more convenient in use. The TRE process of making successful crystals was communicated in detail to BTH, who copied it exactly as regards points (a), (b) and (c) above, as well as in the size and nature of the cat's whisker. Early BTH crystals suffered from a lack of rigidity of the ebonite insulator and the metal fixings to its ends. These difficulties were cured by BTH using a ceramic insulator and properly sticking in the ends. They also did work on samples of silicon which were found to be non-uniform and introduced various time saving devices such as copper plating the silicon and the cat's whisker for soldering. This crystal is essentially the capsule produced by BTH and now in general use.

5. TRE also introduced later (1942) the capacity test for sorting out crystals of good mixer performance.

6. None of TRE contributions were patented, and we have not seen any patents by BTH or others covering the above processes for making crystals.

7. GEC subsequently developed and patented the type of crystal known as 'high burn-out'. The main difference between this and the BTH crystal is in the silicon itself. The processes (a), (b) and (c) are still used, though (b) is less essential owing to greater uniformity of the silicon surface. The cat's whisker has remained the same. The GEC crystal is not an improvement on the BTH type except for its higher burn-out current. Though this is valuable, it remains to be seen whether the GEC type will replace the BTH type except for special applications, on account of its greater cost.

8. As is well known, both BTH and GEC crystals were sent completely developed to the USA. The former is the basis of the Bell Laboratory design which is generally used there.

H.W.B.S.

The crystal mixers made and tested by Skinner at TRE Swanage in late 1940/early 1941 comprised a pellet of silicon soldered to one wire connection and a tungsten whisker about 0.15 mm diameter soldered to the other, with the wires waxed into the open ends of a short length of quarter inch glass tube*. Inevitably, there was a great variation in performance, even when the whisker was moved repeatedly over the surface of the pellet to find the best spot, and pressure was applied to maintain a good mechanical contact while the wax was hardening (some units completely failed to rectify). Partially successful attempts were made by Signal School to solve this problem by sealing the connections through the closed ends of a glass tube. Although crystals made by Skinner at TRE were used in the early experimental Type 271 radar equipments designed and built by Signal School in the early summer of 1941, it was quite clear that a new type of assembly was essential if crystal mixers were to operate successfully outside a laboratory environment.

*Early in 1940, B.J. O'Kane and G.C. Edwards of GEC developed and used a silicon-tungsten catswhisker crystal in a coaxial line system as part of their work on 25 cm AI. The silicon used by Skinner came from GEC.

4.4.3 Development of the Mixer Crystal Assembly

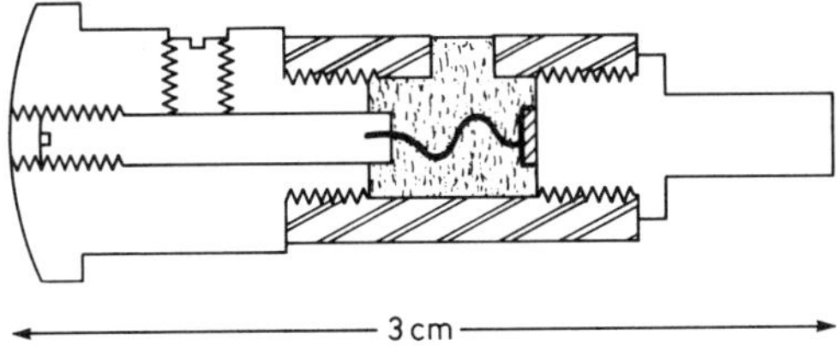

The capsule designed by T.H. Kinman* of BTH, and adopted for use in British radar equipment in April 1941, comprised a ceramic tube about half an inch long with internal threads at each end which located the brass end pieces carrying the silicon wafer and the 'cat's whisker'. The end piece carrying the wafer had an end cap and a pin at its outer end for connecting the mixer to the inner conductor of a coaxial line or to a matching element in a waveguide. The other, which was used to make contact with a connector in the wall of the coaxial line or the waveguide, from which the IF output was taken to the pre-amplifier, was a short cylinder slightly larger in diameter than the ceramic, through which passed a screwed rod carrying the whisker, which was crimped to provide some compliance as contact was made with the wafer by screwing in the rod, which was then locked in position by a grub screw. The end of the whisker was finished by grinding to provide a flat area of contact. Wax could be introduced through a hole in the wall of the cylinder to prevent movement of the whisker caused by dropping or vibration in operational use. The BTH design was adopted soon after by US manufacturers, starting with BTL/Western Electric, except that the silicon wafer and whisker connections were reversed. [A few years later, the design was modified so that it was hermetically sealed. The whisker was then held in place by a thin layer of methacrylate varnish polymerised by gentle heat after the contact was formed].

*From October 1940, Kinman was in charge of development of the crystal receiver module of the experimental GL3 gun-laying radar being designed and built by BTH at Rugby, and was in contact with Oliphant's group at Birmingham, and through them with Skinner. As part of the GL3 programme, Kinman made a small quantity of mixer crystals using an ebonite tube as the body, but these were not mechanically stable.

'Tapping', which was introduced by Skinner early in 1941, was used to 'form' a good rectifying contact after the whisker was set down on the wafer. It was rumoured that a 'standard lead pencil' was used to strike the crystal while its 'back to front' ratio was measured by an Avometer. [For many years, Avometers were the most widely used instruments for measuring voltage, current and resistance in electronic circuits.] Operating on the 100,000 ohm range, the Avometer applied 1.5 V to the crystal rectifier. Experience had shown that crystals would operate efficiently as mixers if their forward resistance as measured by the Avo was less than 250 ohms and their back resistance greater than 2500 ohms. 'Tapping' was repeated until the required back to front characteristic was obtained, or until (more rarely) it was obvious that it would not be.

By May 1941, BTH were making mixer crystals in quantity. In September, production was at the rate of more than 400 per week, of which 80% were suitable for use in GL3, Type271, and other S-band radar equipment. GEC were also producing in quantity, using a different fabrication technique to make capsules interchangeable with the BTH design.

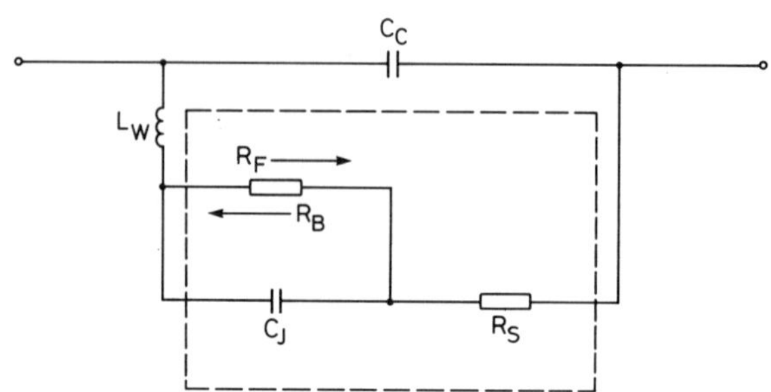

Crystal capsule - equivalent circuit

Although the 'back to front' test identified 'good' crystals, it could not be used alone as a production test of their efficiency at microwave frequencies. A simple theoretical analysis shows that this is inversely related to the self capacitance C_J of the rectifying junction and the total series resistance R_S of the wafer and whisker. For this reason, the 'good back to front' crystals were subjected to a 'CC' test, in which the capacitance was measured by their being connected in parallel with a 100 MHz resonant circuit loosely coupled to a source which caused them to generate 200 microamperes DC output. The reduction in circuit capacitance to maintain resonance was used as a criterion of performance, based upon an empirical correlation of its maximum permissible value with the performance of 'standard' crystals tested in a 'standard' receiver and in an S-band microwave test set developed by the Clarendon Laboratory, who became the accepted British authority on measurement of microwave receiver performance. Although the 'CC' test was effective in selecting crystals for use in S-band, it could not be used effectively to select them for X-band applications, and so they were subjected to a further test in which they were inserted into a waveguide system designed to present a matched load to an X-band source providing a fixed output of a few milliwatts. The rectified current was used as a criterion of performance, being also correlated with the performance of 'standard' crystals in receivers and Clarendon test gear. This 'matched input' test was particularly appropriate for testing CV253, an X-band crystal selected for fitting in radar receivers without adjustment of the matching elements in the waveguide system.

4.4.4 Metallurgy – Sensitivity and Burn Out

In parallel with the development of the crystal capsule at BTH, J.W. Ryde of GEC made an extensive study of the metallurgy of the silicon wafer with the object of improving the sensitivity and uniformity of the rectifying contact, and increasing its resistance to damage by RF power fed into the receiver waveguide from the local or nearby transmitters of a twin mirror system, or leaking through the TR cell of a 'common T and R' equipment.

It was found that sensitivity could be improved by reducing the impurities in commercial silicon powder* to less than 0.01% calcium and iron oxide and 0.04% alumina, and then adding 0.25% aluminium to reduce the resistivity of the bulk material. [That of pure silicon was too high for efficient rectification.] A pure 'aluminium doped' ingot was produced by melting the powder in vacuo in a beryllia crucible. [It was later found that small traces of beryllia had a beneficial effect on performance.] Wafers cut from the ingot and highly polished gave a consistently good performance as point contact rectifiers over the whole surface. As part of the production process, the wafer was cut into small pellets suitable for assembly into the capsule and etched in hydrofluoric acid to remove the surface layer. A new surface layer was then produced by heating the pellets in air at 1000°C, followed by a second etching to reduce the thickness of the surface layer to that required for formation of a good rectifying contact.

*In July 1941, BTH tested BTL crystals against their own, and found that the American crystals were more uniform and superior in performance. They attributed this to the better quality of the du Pont silicon used by BTL. In 1944, BTH requested permission to import du Pont silicon for the production of high burn-out X-band crystals such as CV253, it having been found that the yield when using British material was unacceptably low.

Crystal elements produced in this way were relatively uniform, had a high 'back to front' ratio, and were resistant to damage by RF pulses. GEC were producing 'high burn out' crystal capsules in small quantities by September 1941, soon followed by BTH. An incidental advantage of the new process was that only a few light taps were required to form a good and mechanically stable rectifying contact. Later on, it was found that an unpolished wafer, lightly etched to remove grosser impurities, was uniformly sensitive over its surface, and that use of a molybdenum-tungsten alloy for the whisker resulted in a higher back to front ratio as the forward resistance was reduced by tapping. The mechanical stability of the junction was also improved.

4.4.5 Measurement of Resistance to Burn Out

In early 1942, when GEC and BTH were making high burn out crystals in quantity, it was not possible to assess them during production test by applying pulses of RF power with the appropriate energy content. They were therefore subjected to an arbitrary test in which a 50 Hz alternating voltage across the crystal was increased smoothly from zero to 4.5 V over a period of 4 to 5 seconds and then quickly reduced to zero. If the back resistance was not decreased significantly, the crystal was classed as high burn out.

In 1943, the Radiation Laboratory at MIT in Boston showed that the leakage through a TR cell is in two parts – a 'spike' of about 10^{-9} second duration, followed by a 'flat' lasting for the duration of the transmitted pulse. The power in the 'flat' is too small to cause deterioration of crystal performance, but the thermal time constants of individual junctions are such that a fraction of an erg of RF spike energy can cause serious damage. Thus crystals must be assessed for proof against burn out by subjecting them to a test simulating this effect. In the equipment developed by BTH, a short length of charged coaxial cable was used to apply a series of DC pulses with an energy content of a fraction of an erg which were applied to the crystal by a high speed relay contact. When sufficient energy is applied to cause 'burn out', the back resistance falls below 1000 ohms within a few seconds. Crystals were graded as 'normal', 'high burn out', and 'extra high burn out' by adjustment of the voltage applied to the coaxial line.

4.4.6 The Finished Product

The history of the development of microwave crystals was not one of changes in design, but of continuous study and refinement of fabrication and testing techniques. In the years 1941 to 1944, many thousands were manufactured by BTH and GEC, but however the processes were improved, the output from production was selected rather than passed as good by the test procedures. The yield of 'high burn out' and 'X-band' crystals increased significantly, so that they were almost the norm, with some 'normal' crystals selected for high sensitivity and reserved for use in 'pre-plumbed' X-band receiver waveguide systems, for example CV253, a selected CV111.

Thus a family of devices was built up with an S-band range CV101, 102 and 103 coded according to increased burn out in that order, and an X-band series CV111, 112 and 113, each identified by coloured spots. [The body of the capsule was too small for other markings.] To protect them from accidental damage by incident radiation, each crystal was wrapped in metal foil before shipment.

4.5 Duplexers – TR and TB cells

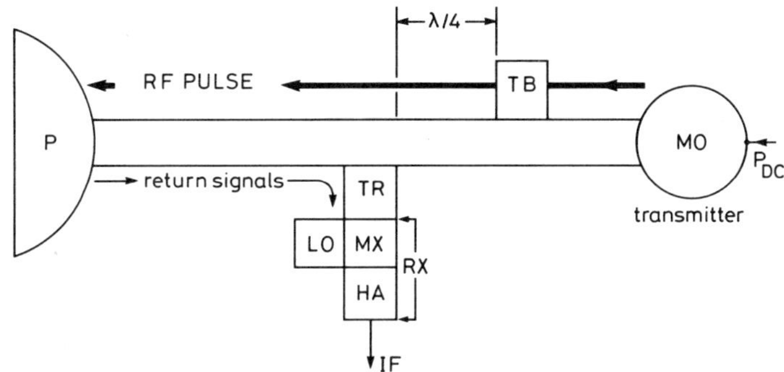

4.5.1 Foreword

In 'common T and R' pulsed microwave radar systems the TR cell and the TB cell share three functions:-

(a) to cause minimum attenuation of the radiated signal when the transmitter is turned on (TB + TR).

(b) to provide adequate protection for the crystal mixer during the pulse, (TR), and

(c) to cause minimum loss of input signals generated by energy reflected from objects within the target area after the RF pulse has passed (TB + TR)

This 'duplexing' function is achieved by filling the resonant circuits of the cells with a gas at low pressure, in which the transmitter power produces a glow discharge and so alters their power transfer characteristics. During the pulse, the glow discharge acts as a high conductance shunting their connections into the waveguide joining the magnetron to the mirror, so fulfilling function (a). [It is convenient to consider the cells as being effectively in series with the connecting waveguide, but equivalent shunt circuits may be used in practice.] In the TR cell, the glow discharge reduces the impedance of the resonator by many times, and limits the power passing through it to the crystal, so fulfilling function (b). [The rhumbatron in a TR cell is a high impedance high 'Q' circuit, and so the damping produced by the low impedance of the glow discharge causes very high attenuation between the input and output couplings.] After the RF pulse has passed and the cells have deionised, signals returning from targets pass through the TR cell to the mixer with very little loss, and are prevented from entering the magnetron by the high impedance of the now passive TB cell. [The quarter wave section between TB and TR transforms the high impedance at TB into a low impedance across the waveguide at the TR entry, so that a returning signal is coupled into the TR cell with very little loss.] This fulfils function (c).*

*In AI MkVII, which operated in S-band and used a coaxial RF connecting system, the TB cell was omitted to save space. The RF circuit was arranged so that the input impedance of the magnetron output circuit simulated the impedance of the missing TB cell. Because of the intrinsically higher losses in the magnetron circuit there was a greater loss of signal than would have occurred with a TB cell, but this was acceptable operationally.

Thus the TR cell must ionise very rapidly when the RF pulse arrives. This requires that there be at that time an adequate number of ions in the rhumbatron gap. These are injected by a 'keep alive' discharge from an electrode sheathed but for its tip, and fed from a negative DC supply which causes it to draw up to 200 microamperes through an immediately adjacent ballast resistor, which prevents relaxation oscillation. The keep alive current is limited to ensure that attenuation of signals passing through the TR cell is the minimum compatible with an adequate supply of ions in the gap. Alternatively, and preferably, the keep alive current can be very low, and a 'pre-pulse' applied to the electrode immediately before the transmitter is turned on, so to ensure that there is a large number of ions in the gap before the RF pulse arrives.

> In 1943, the Radiation Laboratory of MIT showed that the leakage through a TR cell is in two parts, an initial 'spike' with a typical energy content of a fraction of an erg and a duration around 10^{-9} seconds, followed by a 'flat' lasting for the duration of the pulse with a power around 50 mW. Damage to crystal mixers is almost always caused by the 'spike', which can cause intense local heating of the microscopic areas of contact between the whisker and the crystal.

After the pulse has passed, the TR cells must deionise within about 5 microseconds if targets at short range are to be detected with confidence – a longer time can be accepted for the TB cell. Thus the gas filling must have a high recombination rate. Water vapour was used in the early TR cells, but is rapidly cleaned up by the action of the glow discharge which accelerates sorption into the metal parts of duplexer cells, so limiting life unacceptably. Ultimately, it was replaced by mixtures of argon and water vapour, where the water vapour acted primarily as a deionising agent and was not cleaned up so rapidly.

Maintenance of an adequate water vapour pressure, which was the key to obtaining adequate operating lives, was the most difficult problem to solve. It required that the cells be free of contaminants which might remove it by chemical combination, and that the metal parts be processed so that they were in equilibrium with water vapour at the required pressure before the cells were filled with the operating argon mixture. Measurements were difficult and not necessarily reliable. Thus it could be said that the production of TR cells relied more on good cooks than production schedules.

4.5.2 Early Days

The first microwave TR cell was made by A.H. Cooke of the Clarendon Laboratory in mid-1941. Filled with water vapour at a pressure of a few millimetres of mercury, and known as the 'soft Sutton tube', it used the rhumbatron resonator of the NR89 S-band reflex klystron, with diametrically opposite loop couplings for connecting it to concentric lines leading to the mixer and to a T-junction in the line joining the magnetron to the radiating system. It could also be used as a TB cell, and be fitted (with some difficulty) to waveguide systems. To maintain the required background level of ionisation, a radioactive compound was painted on the inside of the glass outside the resonator. This was soon discarded in favour of a DC keep-alive electrode mounted in the tube which had housed the reflector of NR89. The gun tube was retained as a second reservoir for the water vapour filling.

4.5.3 S-band Developments

In July 1941, samples of the Clarendon design were successfully tested by H.W.B. Skinner and A.G. Ward of TRE in their experimental S-band AI equipment. As a result of this, an order for their production was placed with E.K. Cole, who were at that time making NR89 to the Signal School design. They delivered a batch of twelve tubes in September. Thereafter, development of the active element was continued in consultation with Cooke by Skinner and Ward and A.R. Chance of Ekco, with the object of improving life and crystal protection by modifications to the filling and to the production processes. It was roughly one year before satisfactory lives were obtained after a succession of painstaking and small but important changes to the pumping, outgassing and filling processes. Type Approved as CV43 at the end of 1942, the tube was filled with a mixture of argon and water vapour, each with a partial pressure around 6 mm Hg, and operated satisfactorily for a few hundred hours. The second reservoir was reduced to a fraction of its original length to simplify fitting of the tube into duplexer systems. Information on the design and performance of CV43 was sent to USA at the end of 1941.

Because its coupling loops were liable to sparking when operating at power levels over 50 kW peak, CV43 could not be used with the CV56 and CV64 magnetrons which became available at the end of 1941 and were adopted for use in Type271 and AI MkVIII. As a result of this, new resonators were developed by Signal School. These used an apertured flange coupler on the outer wall, which mated with a corresponding aperture in the waveguide. Apart from increasing the power handling capacity to 200 kW peak, use of direct coupling to the waveguide simplified mechanical assembly and setting up procedures, particularly when the cell was operating as a TB. [Because a TB cell can operate efficiently with a lower 'Q' than a TR, it can be coupled more tightly to the waveguide, and be preset to cover the operating range of a particular equipment such as AI MkVIII and H_2S.] Development was completed in mid-1942, and followed by production at E.K. Cole and Type Approval as CV210, CV193 and CV83 in 1943. [the various type numbers refer to frequency variants of the same basic structure.]

With powers above 100 kW peak, lives of the CV210 family of TR cells were limited by deposition of copper on the inside of the glass in the resonator. This arose from the appearance of an 'extra glow' inside the glass near the coupling hole. To overcome this, the TR cell was protected by a pre-TR placed between it and the waveguide joining the magnetron to the mirror. The 1B38 pre-TR, which was developed in USA in late 1943,comprised a quarterwavelength guide capped at each end by a rectangular 'resonant window' of low loss glass sealed into a cobalt nickel iron alloy plate which was hard soldered to the ends of the copper plated steel waveguide. The filling was a mixture of argon and water vapour at a pressure about 1cm Hg. [A similar cell with a single window and a closed end could be used as a TB cell.] In operation, the incident RF power produced a glow discharge which shunted the input window and so reduced the power passing through the cell to a few kilowatts. In its passive state, the cell transmitted low level signals with very little loss. CV294, an alternative design developed by GEC and Signal School at about the same time, used a 'dumb-bell' resonant slot cut into a metal plate which could be mounted between two waveguide coupling flanges. The active element was a thin walled low loss glass tube filled with argon and glass beads, which crossed the gap in the resonator. The dimensions of the assembly were such that it was resonant at the transmitter frequency and so permitted low level signals to pass through with very little loss.

The transmitter pulse produced an electrodeless discharge which partially short circuited the resonant system, so reducing the power reaching the TR cell. The glass beads acted as a de-ionising agent. CV294 could also be used as a TB cell when backed by a shorted quarter-wavelength guide.

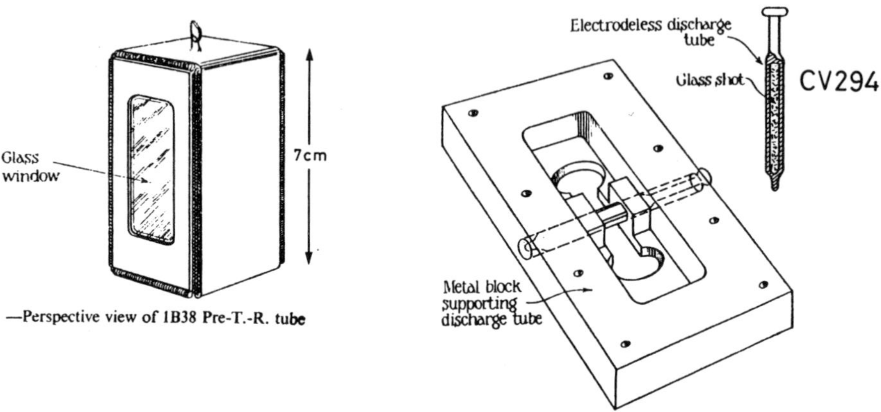

—Perspective view of 1B38 Pre-T.-R. tube

CV.43.

28.5 - 30.0
25.5 max.
76 max.
Fixing Bracket
Note :-
The spacing between the
Reflector Electrode and
the surface of the copper
disc should be between
1/16" and 1/8".
Clamping Ring
1/16" thick.
18.75 ± 0.05.
1/4 max.
Heater Seating
Ring.
62 max.
With the tuner in
a position such that
further unscreening
does not affect the
tuning.
Tuning Plungers 5/8"
dia. 26 T.P.I. Lock-
nuts 23.4 max. A/F.
46 max.
27°
27°
27°
6 B.A. Bolts on a P.C.D.
of 49.2 ± 0.1.
46 max.
27°
27°
38.1 ± 0.1
56 max.
Holes tapped 5/8" dia. 26 T.P.I.
All dimensions in m/m
unless otherwise stated.

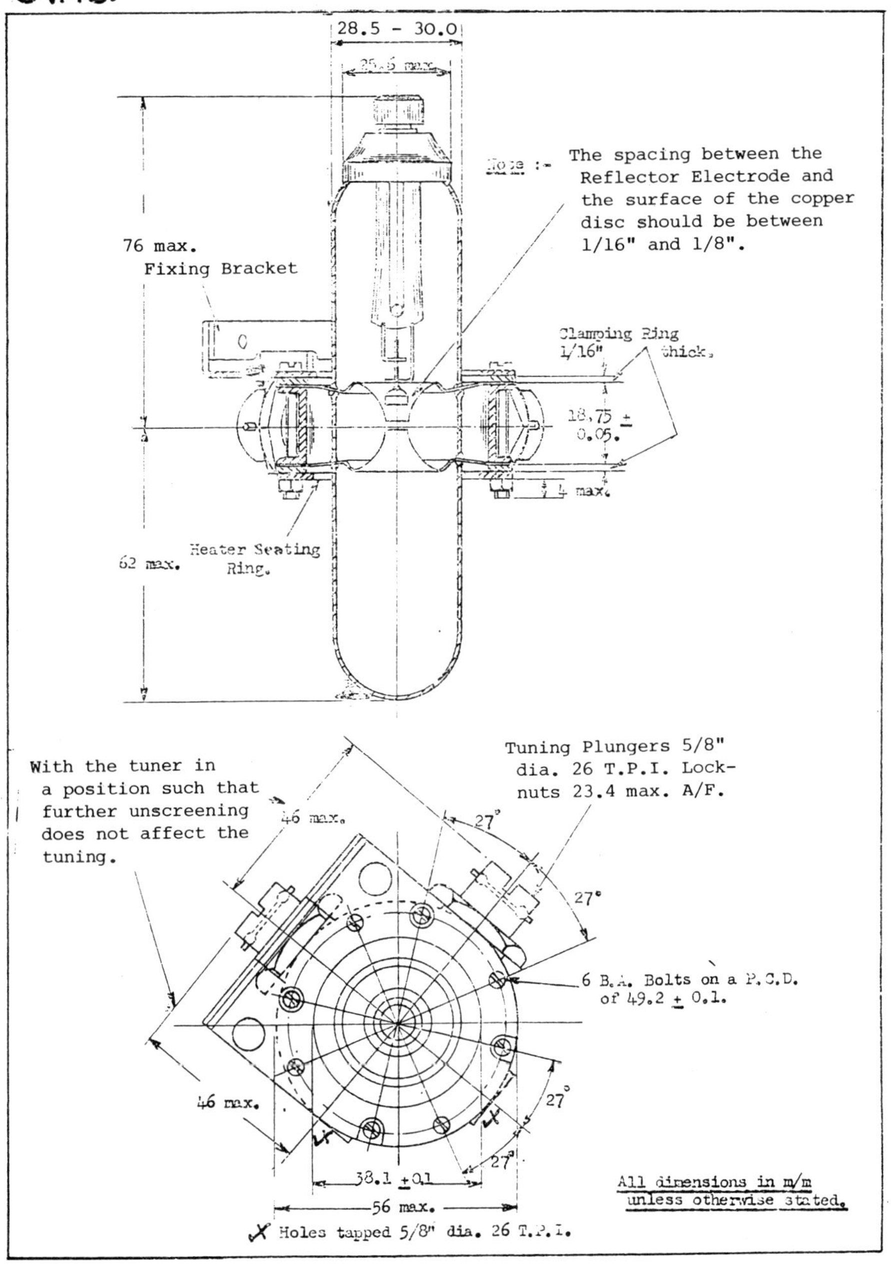

CV 210

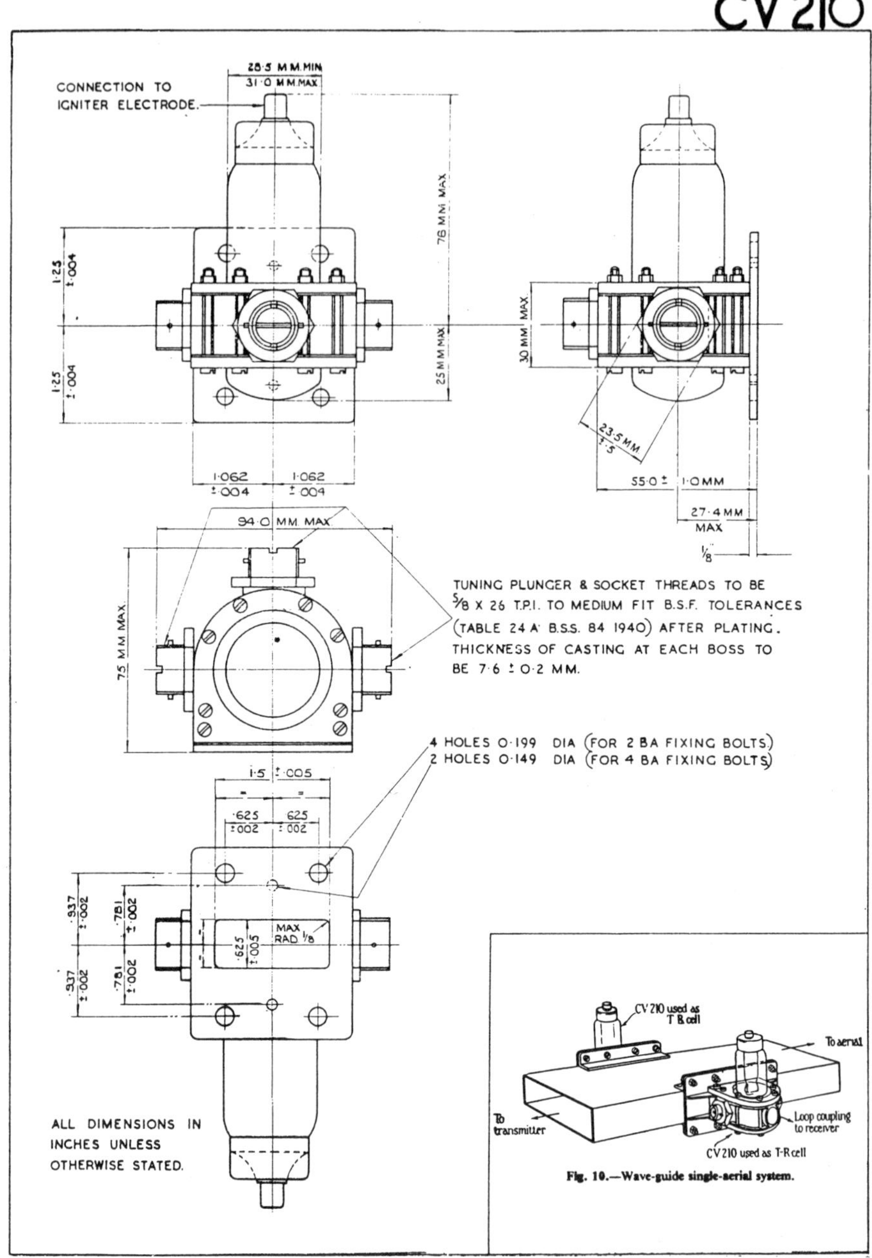

Fig. 10.—Wave-guide single-aerial system.

4.5.4 X-band Developments

In X-band, the design of the TR and TB cells differed from the beginning. The TR cell used a cavity resonator operating in the fundamental mode with a pair of glass windows for coupling in and out. The TB cell was a shorted quarterwavelength guide, closed at its open end by a resonant window in contact with the wall of the waveguide leading to the magnetron.

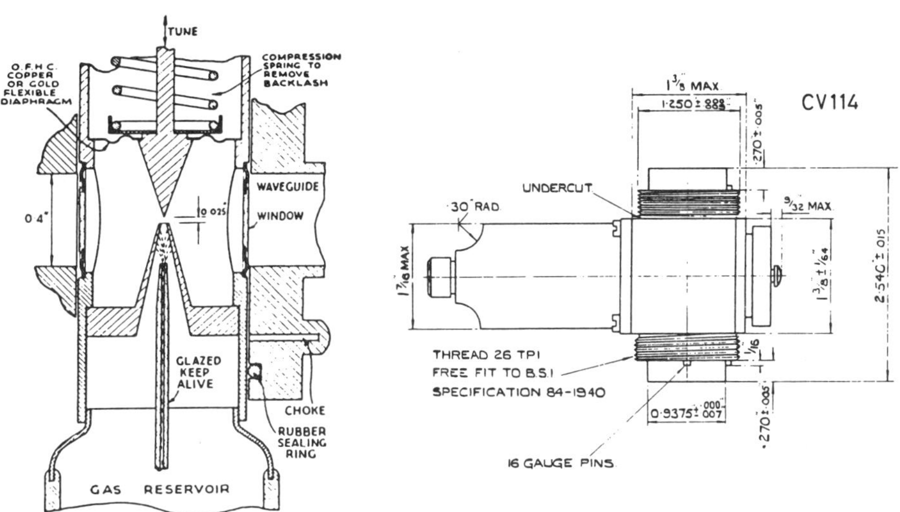

The first X-band TR cell was devised by H.W.B. Skinner and A.G. Ward of TRE in early 1942, and developed by N.L. Harris and his group at GEC. To simplify construction and to meet urgent requirements for production, the resonant cavity was bored cruciformly from a brass block so that circular low loss lead glass windows sealed in copper annuli could be soft soldered at each end. Outside each window, a short length of circular waveguide and a quarterwavelength transformer coupled the cell to the mixer assembly and to the waveguide joining the magnetron to the aerial system. The aperture at the top was closed by a copper diaphragm forming one 'horn' of the resonator, which was pierced to provide entry for ions generated by the 'keep alive' discharge. The vacuum envelope was completed by a flared copper cylinder sealed to a glass bulb which acted as the gas reservoir and also housed the partially sheathed keep alive electrode. The aperture at the bottom was closed by a corrugated copper diaphragm carrying the other horn of the resonator, which could be moved up and down by a differential screw mechanism to tune the cavity over the range 3.1 to 3.3 cm. Both diaphragms, and the reservoir assembly, were soft soldered in position. The early tubes were filled with water vapour, but there were problems with 'clean up' similar to those in early S-band cells, and so it was replaced by a mixture of argon and water vapour at partial pressures around 6 mm Hg. With a keep alive current of 300 microamperes, breakthrough into the mixer was about 0.2 ergs per 'spike' with 30 kW peak incident power. The low level attenuation (insertion loss) averaged 3 dB. Deionisation time was less than one microsecond.

C.V.221.

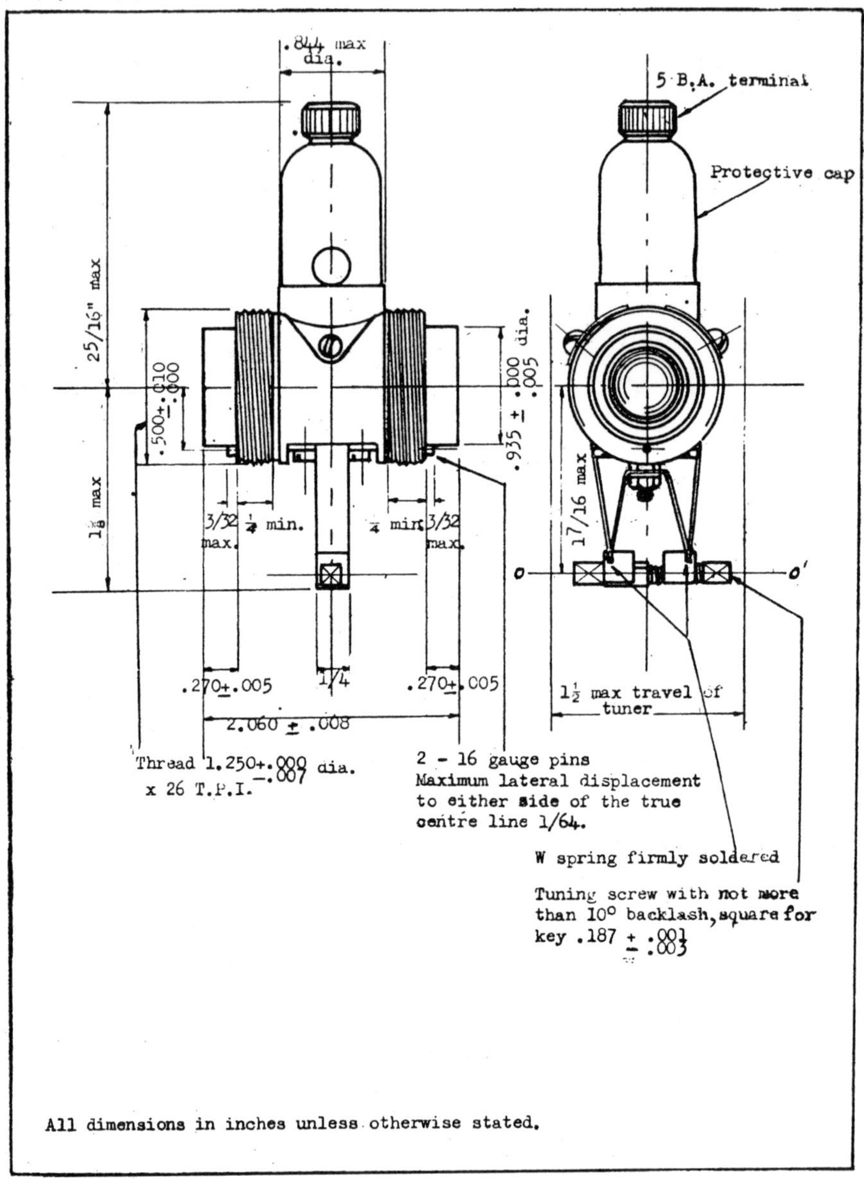

CV114 was being made in quantity by GEC in December 1942. Type Approved for use in AI and H_2S, it continued in production until 1944, when it was superseded by CV221.

Because of its soft soldered construction, CV114 could not be outgassed effectively on the pump before the argon water vapour mixture was introduced, and was limited to small scale production. Its 'spike' breakthrough was high enough to cause a significant deterioration in the performance of X-band mixer crystals. [Although the crystals could withstand spikes with an energy of a fraction of an erg without suffering catastrophic damage, it had been found that 0.05 erg could cause a measurable increase in conversion loss.]

To overcome these deficiencies, GEC developed E1365, beginning in May 1942. It differed from CV114 in having a tellurium copper body and gold ring seals for the window and diaphragm assemblies, so that the whole assembly could be clamped together and baked on the pump at 400°C to outgas the assembly and form the vacuum seals by diffusion of gold into the mating copper surfaces. [This process had been used previously by GEC in the production of magnetrons.] Narrowing the gap between the horns of the resonator reduced spike breakthrough to about 0.03 erg with 45 kW peak incident power and 150 microamperes keep alive current. As a result of this change, the size of the cavity was reduced, and likewise the thickness and diameter of the windows. This resulted in a reduction in low level attenuation to about 1.25 dB. The recovery time increased to a few microseconds, but this was acceptable in operational use. Filled with a mixture of argon at 9 mm Hg and water vapour at 6 mm Hg, E1365 gave lives of over 500 hours with transmitter powers in the range 45-60 kW. Development was completed in early 1943. Type Approved as CV221 in September 1943, it continued in production through 1946.

Although CV114 and CV221 could be used as TB cells, they were unnecessarily complicated and expensive for this function. A much simpler device, E1465,

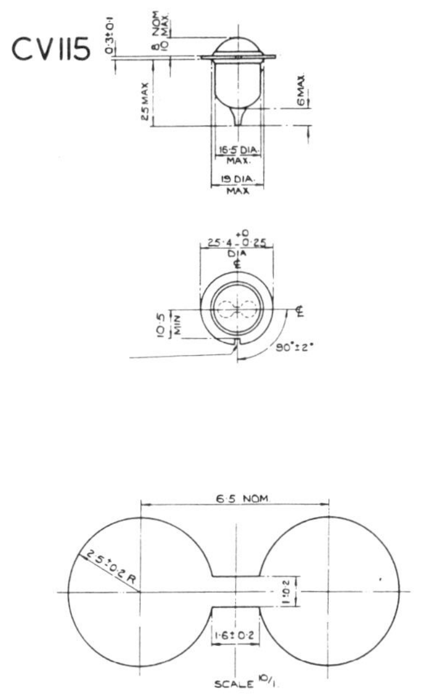

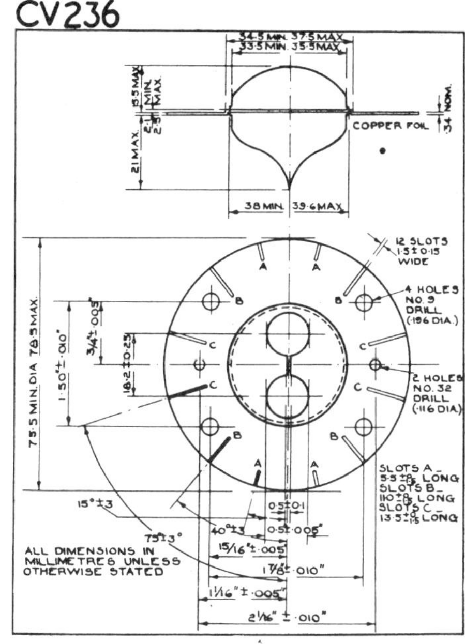

was designed by TRE and GEC in late 1942. This used a 'dumb-bell' resonant slot die-cut in a copper disc, with glass bulbs sealed to each side of the disc to form the gas reservoir. The disc was clamped to the broad face of the waveguide with the smaller bulb facing inwards. The other was covered by a short length of waveguide incorporating a tuning plunger which could be preset so that the assembly would operate over the wavelength range of a particular equipment such as H_2S or AI MkVIII. The filling was argon at a pressure of 700 mm Hg. No water vapour was required because the recovery time could be considerably longer than that of its associated TR cell. Type Approved as CV115 in 1943 it continued in production through 1946. A scaled up version (CV236) was used in some S-band equipment.

By late 1944, British developments had been overtaken by American technology*. US manufacturers were producing 1B24, an X-band cavity type TR cell which could be clamped between a pair of standard waveguide coupling flanges. This was followed by 1B63, a broadband cell which combined in a single unit the functions of a pre-TR and TR cell. This used a threequarter wavelength copper plated steel waveguide, closed at each end by a resonant glass window set in a waveguide coupling flange, and having two low capacitance dumb-bell iris resonators set across the guide about one quarter wavelength apart. Keep-alive electrodes provided a glow discharge fringing the gaps of the resonators. In its passive state, the whole assembly operated as a bandpass filter with about 5% bandwidth.

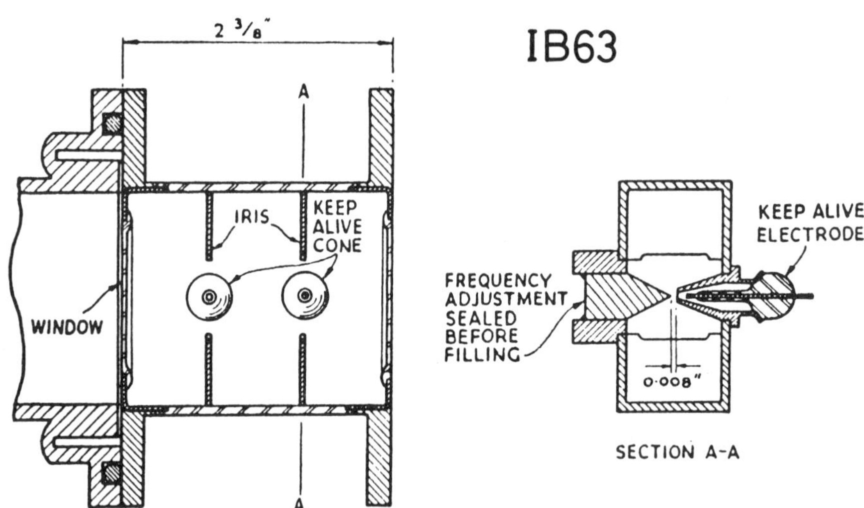

*One reason for this was the invention of 'Kovar', a cobalt nickel iron alloy to which flat rectangular glass windows could be sealed without the need for a surrounding compliance. In X-band the apertures in the Kovar had semicircular ends. Those in S-band had rounded corners. The glass overlapping the alloy on the inside of 1B63 reduced sputtering of metal on to the window by the RF power discharge at the input end. Because of their lower chloride content, American glasses had lower RF loss than their British equivalents.

Chapter 5

Cathode ray tubes for radar displays

5.1 Foreword

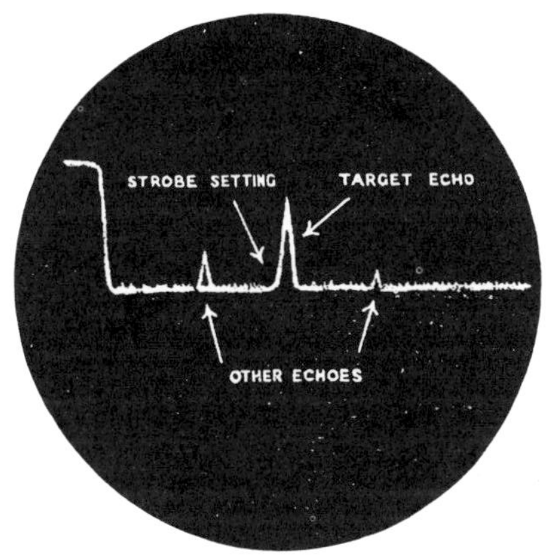

A-scan data display

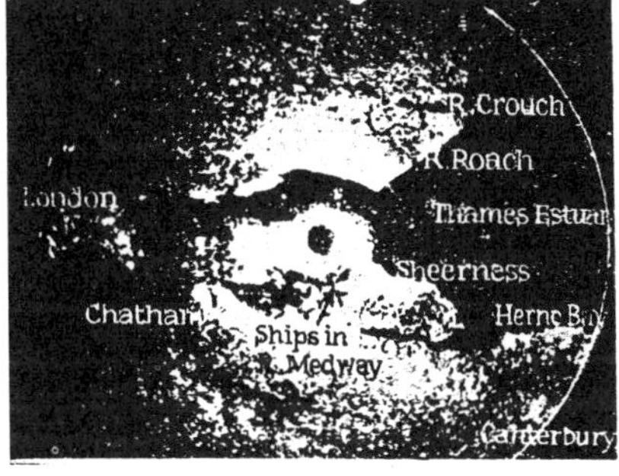

Plan position indicator

Broadly speaking, there are two kinds of radar display:

Those which present data on the range and bearing of objects within the view of the aerial system with which they are associated. Typical examples are the 'A-scan' range displays in Naval and land-based early warning radar, GL MkIII, and in AI and metric ASV.

Plan Position Indicators, which display a 'picture' of the surroundings of the aerial system, which rotates continuously to provide 'all round looking', as in H_2S and ASV, and centimetric Naval radar.

5.2 Data Displays

In general, data displays require that the operator learn how to interpret the information on the screen and to decide what action be taken, particularly when multiple targets are scanned by the aerial. Alternatively, the display may provide data in the form of an instruction to the operator, for example in navigation aids and fire control systems. In a typical A-scan display, the spot on the screen of the electrostatically focussed and deflected cathode ray tube moves from left to right during each pulse interval, returning to its rest position before the next pulse is transmitted. By decreasing in steps the time taken for the scanning voltage to reach its peak amplitude, the length of the trace represents shorter and shorter ranges. The target is indicated by a vertical spike at a position corresponding to its distance from the radar aerial, which may be rotated to and fro to maximise the amplitude of the spike and so give an approximate indication of the target bearing. This can be measured with higher accuracy by observing and bisecting the angle between the half-amplitude settings of the aerial rotation control. [This procedure was used with the Naval Type 279 radar.] Alternatively, and preferably, the beam from the aerial can be split by switching the receiver from one to the other of a pair of aerials directed at equal small angles on either side of the mean position, and simultaneously switching the polarity of the spike voltage applied to the Y-deflector plates of the cathode ray tube, so producing a back-to-back display in which the spikes will be equalised as the aerial is directed towards the target. In some back-to-back displays, the trace is vertical, which makes it easier for the observer to match the spike amplitudes by eye.

[In common T and R systems, the switching mechanism must be capable of handling the transmitter power, and so the radiating element at the focus of the paraboloid mirror may be rocked from side to side to provide the split.]

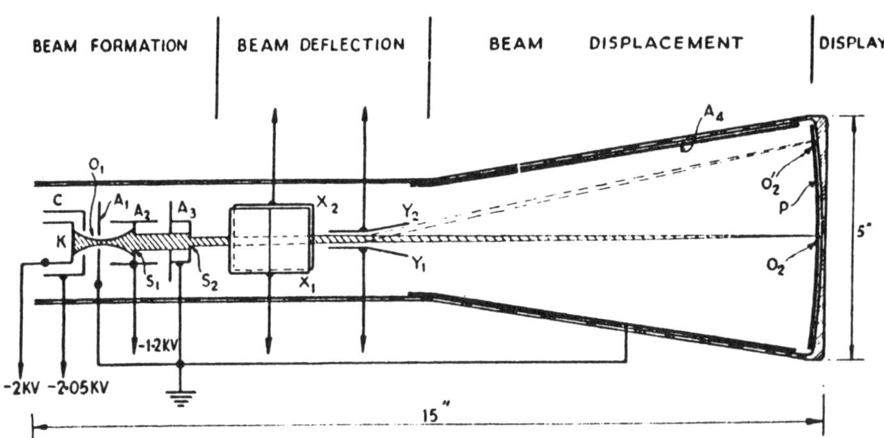

Cathode-ray tube with electrostatic focus and deflection. K—cathode. C—control or Wehnelt electrode. A_1, A_2, A_3—converging lens electrodes. S_1, S_2—beam collimating apertures. X_1, X_2, Y_1, Y_2—deflector plates. A_4—final anode collecting secondary electrons from the phosphor P. P—cathodoluminescent screen. O_1—first focus of the beam. O_2—second focus of the beam.

Because the information on the screen is renewed hundreds of times per second, there is seldom a need for persistence of the image*, and so adequate definition, (determined primarily by the diameter of the electron beam when it strikes the layer of phosphor on the screen), and maximum efficiency of conversion of the energy in the beam into visible output, are prime criteria in the design of the cathode ray tube. The diameter of the beam is determined by the geometry of the focussing electrodes through which it passes on its way to the deflector assembly and thence to the screen. Movement of the focussing electrodes under vibration will cause blurring of the spot on the screen.

*In October 1938, there were discussions between Bawdsey R.S. and GEC concerning the supply of 'AJ' cathode ray tubes. For reasons of security, Bawdsey did not divulge the intended use. (Their prognosis was that tubes with persistence of the image on the screen would provide an anti-jamming capability for CH displays, because signals from targets would 'integrate up' while background noise and other interference would 'average out'. In correspondence between Rowe of Bawdsey, Tizard of Air Ministry and Paterson of GEC in late 1938, reference is made to the urgent need for afterglow phosphors to be developed for AJ tubes, to Merton's proposal for double layer screens (see 5.4) and to an offer by GEC, who were aware of Merton's work, to supply phosphor powders and tubes. These were tested with some success at Bawdsey in April 1939. (This was almost certainly the origin of W.B. Lewis' comment on the use of 'modulation' instead of 'deflection' techniques and the need for phosphors with a shorter afterglow made at the CVD meeting with GEC in September 1939). Following the tests at Bawdsey, L.C. Jesty of GEC was 'allowed in' to RDF, and continued the research on afterglow tubes. One outcome of this work was the development of VCR140(A), which in 1940 was the first PPI tube to be used operationally.

It is an obvious advantage to have high deflection sensitivity so that the power required from the scan and signal voltage amplifiers is kept to the minimum. This requires that the deflector plates be as closely spaced as possible. Movement of the deflector plates caused by vibration will result in unwanted deflection of the spot. Thus ruggedness and rigidity of the beam focussing and deflector assemblies are essential if the tubes are to operate successfully in mobile and aircraft equipment. During the war years, there were major advances in this aspect of design.

Because the scanning and signal waveforms in A-scan displays are similar to those in oscilloscopes used for electrical measurements, the cathode ray tubes in the first CH displays were the same as those developed, manufactured and used by A.C. Cossor Ltd in their commercial oscilloscopes. [L.H. Bedford, who was in charge of research and development at Cossor, and O.S. Puckle, who was a specialist in oscilloscope design, were two of the few engineers outside the Air Ministry group at Orfordness who had been told about RDF. Cossor engineered and built the receivers and displays for the first CH systems and the early GL equipment, and thereafter continued to develop new displays and cathode ray tubes].

5.3 Plan Position Indicators

In PPIs, the blacked out spot moves radially outwards from the centre of the cathode ray tube screen during each pulse interval, returning to the centre before the next pulse is transmitted, and rotating in synchronism with the aerial system. Echoes from objects within the selected operating range of the equipment cause brightening of the spot at a position corresponding with their range and bearing, so painting a 'picture' of fixed and moving objects against the background of sea

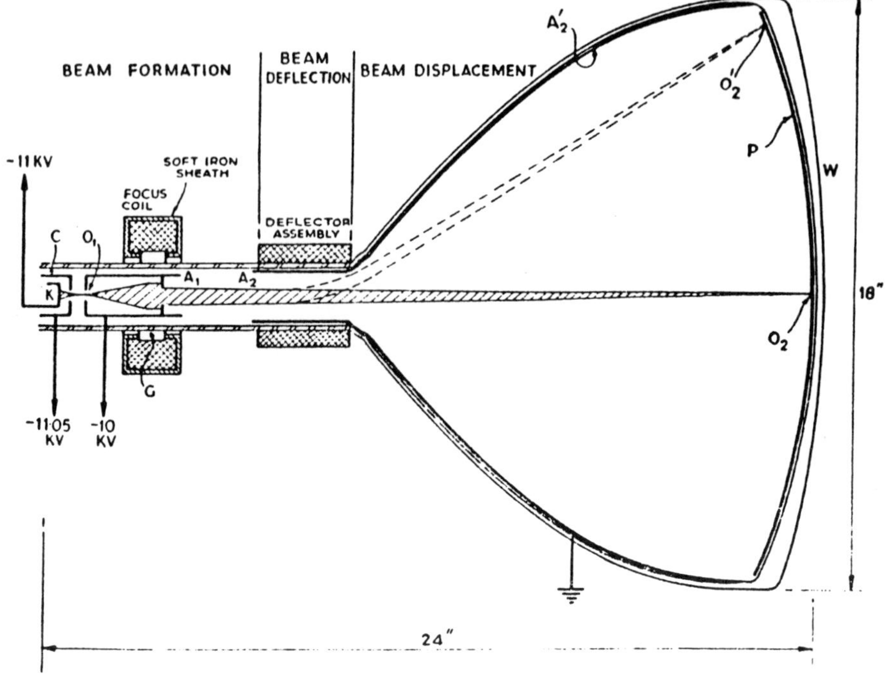

Cathode-ray tube with magnetic focus and deflection.
K—cathode. C—control electrode, also used for beam current modulation.
A_1—first anode. A_2A_2'—final anode. G—gap in focus coil magnetic shield.
O_1—first focus of the beam. O_2—second focus of the beam.

or sky surrounding the aerial system, for example in Naval and land based early warning radar, and in airfield surveillance systems. The rotating aerial in the H_2S airborne radar looks downward and slightly forward, so that the PPI displays strong signals from hard targets such as buildings, railways and roads against the background of smaller signals from the ground below the aircraft and absence of signals from lakes, rivers and the sea. The range is selected by decreasing the time taken for the radial scanning voltage to reach its peak amplitude, so that the picture on the screen represents correspondingly smaller areas surrounding the aerial system. The screen of the cathode ray tube must retain the images of targets for more than one rotation of the trace, so that the observer has before him at all times a useful representation of the target area. Thus the phosphors used in the tubes must give adequate brightness in the afterglow which persists after the beam has passed, and retain the image for long enough to provide an acceptable picture, but not so long as to cause masking of new images by those recorded earlier, for example when the 'heading' of the display is changed as a result of an alteration in course by the ship or other vehicle in which the radar is installed. [Some shipborne installations are arranged so that the display is always 'north heading', using a servo system to control the deflector system and so compensate for changes of course].

Because the rate of rotation of the aerial system may be as low as 2 rpm in land based installations and up to 60 rpm in aircraft equipment, a prime factor in the design and production of PPI tubes is the choice of phosphor and close control of its chemical composition and the thickness and uniformity of its deposition on the inside of the tube face.

PPIs are 'picture' tubes, and so they require little interpretation by the observer, who can relate what he sees to maps of the terrain over which the aircraft is flying, or use the information to study the disposition and courses of a number of ships including submarines at sea or aircraft approaching or leaving an airfield or a combat zone. Compared with A-scans and similar systems, PPIs have the advantage that they can display simultaneously the range and bearing of many fixed and moving objects. Range can be measured by reference to marker rings engraved on the window in front of the screen, which also acts as a filter diminishing the flash which precedes the afterglow, or to rings produced on the screen by controlled modulation of the electron beam. Bearings can be measured by means of a cursor line engraved on the window, which is rotated with reference to a protractor scale surrounding the display. VCR140 and VCR516, the first PPI tubes, were magnetically focussed and deflected, and used a pair of deflector coils which were rotated round the neck of the tube in synchronism with the aerial system by a servo motor in the display unit. Later displays used two pairs of fixed deflector coils set at right angles to each other (as on TV tubes). These were driven by power amplifiers deriving their input signals from a rotary transformer in which two perpendicular pairs of stator windings were coupled to the servo driven rotor. A similar system was used in H_2S, first with VCR97, an all electrostatic tube, and then with VCR530, which used electrostatic focussing and magnetic deflection.

Apart from the fundamental difference in screen characteristics, the design of PPI tubes has much in common with that of TV tubes, e.g. wide angle scanning in both X- and Y-directions, and so the first Naval and land based PPIs, which came into use in 1940, used 12 inch Cossor TV tubes. After a short time these were modified to incorporate a double layer phosphor screen developed specifically for this purpose. (see Table 1). The design remained substantially unaltered for most of the duration of the war, apart from minor improvements in fabrication techniques and an increase in the length of the neck to accommodate longer deflection coils and so improve deflection sensitivity. [VCR530, an electrostatically focussed and magnetically deflected tube, which was developed for use in H_2S came into service in 1944 and was then used in relatively small numbers.]

Because of limitations in time, and on the space and power available for the deflection system and its associated power amplifiers, the PPI in H_2S was a modified VCR97, a 6″ all electrostatic tube which was already being manufactured in quantity for A-scan and other data displays. Used operationally in 1942, VCR517 operated with a higher beam current to increase luminous output and consequently had a slightly larger but acceptable spot diameter.To simplify design of the deflection circuits, the dimensions of the X- and Y-plates were modified to equalise their deflection sensitivities. The phosphor was a mixture developed specifically for use in displays rotating at around 60 rpm.

Extract from a report on:

Trials of 9″ P.P.I. in H.M.S. Watchman, 23/7/42 – 3/8/42'

[The PPI display was fitted to a Type 271 10 cm radar system]

Preliminary trials were arranged in the vicinity of Devonport Harbour after the ship had carried out D.G. trials and these were proceeded with on the 27th July. During the D.G. trials careful note was made of the possible effect of the D.G. coils on PPI but no apparent effect was noticed, centring of the trace and radial sweep remaining perfectly aligned. (In connection with this it should be noted that during the subsequent correction for permanent magnetism of the ship in Devonport Dockyard, the intense field to which the ship was subjected completely misaligned the trace on PPI.) After the D.G. trials were completed, the aircraft only made one approach run at about 1,000 ft. – this was not observed on PPI, although the aircraft was plotted from 4 miles by Type 286. It was, however, possible to fix the ship's position and indicate the approach to the swept channel by means of PPI as the weather had thickened and the navigator was uncertain of his position. Accepting the bearing and ranges of Eddystone and Drakes Island as shown by PPI, the ship was turned and proceeded into harbour at 20 knots, land not being optically visible until the boom defence ship was reached.
[D.G. – 'Degaussing']

The ship left for Londonderry at 1530 hrs. on July 30th. During the initial stages of the journey various ships and land echoes were plotted, the maximum seen being high points in the vicinity of Redruth at 85,000 yds. Tests were carried out to ascertain if it was possible to upset the alignment of the M type receiver on the turning mechanism (10° steps) but this could not be done. The maximum turning rate that could be accomplished with extreme effort was approximately 7 r.p.m.

A submarine awash on the surface was plotted out to 8,600 yds. (It was not seen approaching the ship as 274 was temporarily shut down in an attempt to trace rather frequent frequency drift of the Sutton tube). At 2200 hrs. when attempting to move PPI from 271 office to the Bridge position it was discovered that the tube H.T. transformer was leaking impregnation and subsequently it developed a complete breakdown of insulation.

This was traced to a faulty H.T. condenser, but attempts to get PPI working after replacing the condenser were not successful, the unit remaining out of action for the remainder of the trip to Londonderry.

A signal had been made to Eastney for a replacement and a day was spent in replacing the H.T. transformer and generally checking over the unit, replacing it in the ship and testing late the same day. The ship sailed at 2000 hrs. to rendezvous with an outgoing convoy at 6 a.m. the following day. The convoy was first sighted on PPI at 27,500 yds. and at this range it was only possible to distinguish echoes in the 'noise' on the 271 indicator unit with careful searching, and had PPI not shown their presence, it is extremely doubtful that they would have been noticed at all. At 25,000 yds. it was possible on PPI to distinguish individual targets showing the presence of a group of vessels but on the 271 indicator unit, it was still hard to see echoes without very careful searching. It was noticed, however, that at extreme ranges the rate of sweep of the aerials had to be slowed down to about 2 r.p.m. to sufficiently energise the 'after-glow' screen, as random 'noise' tends to energise the screen to a certain level, making differentiation of small echoes uncertain when relying upon the 'after-glow' characteristics of the screen alone. (See note 6). It is possible under these conditions to see the small signal on the scan line as it sweeps through the target bearing even at higher rotating speeds.

At approximately 18,000 yds. it was easily possible to identify 80% of the ships as separate targets, and going astern to the convoy it showed the separate lanes in which the group of vessels were split up. Under these conditions any new target appearing in between the lanes could have been easily seen.

Summary and General Observations

1. It would appear that PPI can be of immense value to escort vessels, but as size is important the later models of 9″ Air Ministry PPI will be a problem to fit:- it may be necessary to re-design the unit.
2. Training methods:- In extreme cases hand training (if attention is paid to pedestal design in order to keep the necessary torque as low as possible) can be used satisfactorily without imposing undue fatigue on the operator. Some method of power training is, however, considered desirable, with continuous rotation, and provision for arc sweeping.
   [Training – rotation of aerial system]
3. There appears to be no doubt that PPI improves the tactical value of 271 in as much as small echoes are far more readily observed than with the Indicator Unit.
4. Mental strain on operators is considerably less with PPI than with the normal Indicator Unit, in spite of the increased physical strain imposed by the necessity of continual hand sweeping. In connection with this point all operators on Watchman were closely questioned as to their preference and they were unanimous that with PPI they felt that an hour on watch would not be excessive, whereas with the Indicator half an hour was more than most could do without strain. They all considered after handling PPI that there was no possibility of missing a small echo with considerably less strain and, in addition, as they were presented with a 'Plan' it gave considerably more interest to their job. The 'Plan Position' also enabled them to identify previous targets much more easily and in one practice watch with one operator he was easily able to identify eleven known targets. This operator remained on watch for one hour and 40 minutes and relinquished his post quite fresh and willing to continue.
5. Ranges. The continuous range adjustment fitted seems to be unnecessary and in practice two ranges seem to be indicated: one short, of 0–15,000 yds. and one long, of approximately 0–100,000 yds. In addition, there appears to be no reason for retaining the 'linearity' control as an adjustable feature. The scan can be set up initially and left.
6. Differentiation of individual targets in close proximity. The usefulness of PPI is closely bound up with this point, especially when used by convoy escort vessels. The unit installed on 'Watchman' could have been more useful had it been able to differentiate ships in the convoy at small ranges when

individual ships were in close proximity to each other. Since this is a function of video amplifier and limiter frequency response, close attention must be paid to future design of these particular parts of the equipment, and signal/noise ratio maintained at a high percentage.

7. Scale illumination. The bearing scale of PPI must be clearly lit. In Watchman this was provided for by external dim lighting available by push button control but difficulty was experienced with this arrangement, and some other form of illumination must be used in future models.

8. There seems to be no need for additional scales showing either ship's head or relative bearings since the fitting of the bearing indicator (A.P.W.5212) removes the necessity for this.

9. Finally, as a certain amount of valuable time was lost on Watchman due to the failure of the H.T. transformer, it would be highly preferable to install a second equipment in another vessel and complete the trials not carried out in the first installation.

5.4 The pattern of Developments

In the words of three of the senior engineers* whose groups were responsible for the design of cathode ray tubes for a wide variety of applications,

'The war years have seen steady progress in cathode ray tube techniques. Whilst no revolutionary development has occurred, the overall improvement resulting from the summation of a series of relatively minor advances has been quite considerable – important advances have been made in regard to our knowledge of cathode loading [the electron emission per unit area],which has been correlated with operating life – it is now [in 1946] possible to predict electron gun performance [how the geometry of the cathode and its associated electrodes affects the beam current and the size of the spot on the screen] with engineering accuracy. By far the greatest progress has been made in the domain of afterglow screens – a practical result of this work has been a steady improvement in the uniformity of screen characteristics.'

> *L.C. Jesty, GEC Research Laboraratories; H. Moss, A.C. Cossor Ltd; R. Puleston, EMI Research Laboratories (writing in J.I.E.E. Vol 93 Pt111A, 1946).

The development of screen materials and the techniques for their deposition and assessment was a lengthy and laborious process. Notable contributions were made by L.C. Jesty, J. Sharpe and R.G. Hopkinson of GEC, by R. Puleston and S.T. Henderson of EMI, and by G.F.J. Garlick of TRE (working at Birmingham University). Their work made possible full realisation of the potential value of Plan Position Indicators, which were later used in large numbers for commercial air traffic control and marine radar. A few years after the war, the double layer and mixture screens were replaced by fluorides which produce an orange afterglow without a preliminary flash of very high intensity. Ultimately, the function of the afterglow screen in airfield surveillance was replaced by a computer data store and a high brightness three colour tube for daylight viewing.

[The use of three colours made possible the identification and labelling of large numbers of aircraft approaching an airport.]

'The great majority of the research [and development] work during the war years was directed towards electrostatic tubes – magnetic tubes which were used during the war years were (apart from the screens) merely of the 1939 television type, except for minor changes, and little effort was expended on improving them until the very end of the war.'

By that time, development of VCR530 and CV254 was complete and they were being used in small numbers in H_2S and Naval radar. They were electrostatically focussed and magnetically deflected, and combined the ease of plug-in replacement of electrostatic tubes with the wide angle of deflection of magnetic types. To reduce the power required for deflection, they were fitted with deflector coils twice as long as those used with comparable TV tubes. Sensitivity was increased further by their longer necks, which increased the distance from the centre of the coil to the screen.

During the war years, over sixty different cathode ray tubes were designed and manufactured in quantities from several hundred to over ten thousand. A considerable number were variants of the twenty six main designs, or were specified and tested for operation at different voltages to meet the requirements of individual types of equipment. Of the twenty six designs, fourteen were electrostatically focussed and deflected tubes for A-scan and other data displays, and eight were PPIs. Although they worked in close collaboration, each manufacturer used his own fabrication and assembly techniques for design and construction of the electrode systems, and so exchange of information was limited to the fundamentals of design. In many cases, an individual tube specification would be met by more than one manufacturer, each using his own in-house technology to make tubes which were direct plug-in replacements for each other. This not only provided second sourcing to meet production requirements and an insurance against loss of either factory, but also protected commercial security and initiative in what was and would again be a highly competitive market. Apart from the work on phosphors at Birmingham University, and development of the 'Skiatron' (see 5.5) by R.A.R. Tricker and other members of the Signal School group at Bristol, all the research and development was done at contractors' laboratories and coordinated through CCRTD, the CVD Committee for Cathode Ray Tube Development, which had roughly equal numbers of members from Industry and the Establishments. Recalling the senior engineers' comments, it could be said that their success was based on highly developed bespoke engineering, in which one each of eighteen different bulbs, eighteen electron gun and deflector assemblies [based on known technology but improved and ruggedised as a result of the free interchange of basic information], and eight phosphor screens [most of them improved or developed to meet new operational requirements] were combined to meet the individual requirements of equipment designers (see Table 1). [Before the war the most commonly used phosphors were 'television white', and 'willemite green'(a zinc orthosilicate used in almost all oscilloscopes and monitoring tubes).]

Like the pre-war oscilloscope tubes on which their design was based, all the data display tubes were electrostatically focussed and deflected. As a matter of urgency, the electrode assemblies had to be re-engineered to provide immunity from microphonicity and to make them rugged enough for use in aircraft and military vehicles. VCR97 was the first tube to be designed specifically for airborne applications. Used thereafter as a general purpose indicator, there were about ten variants of the basic design, including VCR138, which used the same neck and electrode assembly attached to a smaller bulb, and VCR517, used as the PPI in H_2S, which was designed to give a higher beam current and so more luminous output from the afterglow screen which replaced the willemite used in VCR97. Many thousands were manufactured.

An important innovation was the use of post deflector acceleration (PDA) in which the beam is accelerated as it travels from the deflector assembly towards the screen. The PDA electrode was an annular coating of graphite on the inside of the bulb near the screen, to which contact was made by a 'button' connector sealed into the bulb. [This technology originated in USA.] The use of PDA made possible the use of higher beam voltages and so the production of much higher luminous output from the screen, which in tubes without PDA was limited by flashover between connections in the base, particularly at high altitude. [In tubes without PDA, the beam voltage is limited by the negative potential which can be safely applied to the cathode (the final anode of the electron gun and the deflector assembly are at earth potential). With PDA, a positive potential is applied to the accelerator to increase the beam voltage by a factor of two or more without risk of flash over]. British manufacturers also adopted American technology for bulb manufacture, in which a moulded face plate is sealed to a neck and flare. [Up to that time, almost all British cathode ray tubes used bulbs blown into a mould]. This resulted in a much lower curvature of the face of the tube and a significant improvement in its optical quality.

The afterglow screen in the first PPI tubes (VCR140, VCR516) was a double layer system in which the electron beam produces a short 'flash' of blue light from the layer of phosphor on which it impinges. This is absorbed by the second layer, which faces the observer, and there generates a yellow afterglow which is viewed through a colour filter so that the flash is not unacceptably intrusive. Screens of this type had been proposed by G.H. Merton in 1936. At about the same time E.G. Bowen had suggested the use of a PPI in early warning radar, but development of a successful double layer screen required more than two years of experiments with different phosphors, their activators*, and the thickness of each layer. The build up and decay times of the double layer phosphor were too long for it to be used in the H_2S Plan Position Indicator, which operated at around 60 rpm (compared with 2–10 rpm for the earlier radar systems). [At a CVD meeting in September 1939, W.B.Lewis of the Air Ministry group at Bawdsey had spoken of the use of 'modulation' techniques for radar displays, saying that they gave a better response to small signals than deflection of the beam, but that realisation of the full potential the modulation technique would require a phosphor with a shorter build up and afterglow than those then available]. In 1942, EMI produced the AB mixture phosphor, a blend of sulphides and their activators* which provided build up and decay characteristics which met the H_2S requirements. In the mixture screen, the blue light generated by the electron beam is absorbed within the crystal structure, so reducing the initial flash.

*Activators are the metals which are added in minute quantity to the sulphides, silicates and other crystals from which the phosphor is prepared. They affect both the colour and the decay time of the light emitted from the screen, which can also be affected by small changes in the processes used for deposition of the screen.

5.5 The Skiatron

The skiatron was invented pre-war by A.H. Rosenthal of Scophony Ltd, and developed by R.A.R. Tricker, P.G.R. King and J.F. Gittins of the Signal School group at Bristol to meet requirements for a Plan Position Indicator which could be viewed by a number of observers or combined with a plotting table. Unlike afterglow tubes, the skiatron is a dark trace tube in which the electron beam produces a persistent image against the white background of the screen. Developed by the end of 1942, and manufactured in small quantities by Cinema Television Ltd they were available for use in shipborne and airfield plotting table displays in 1943. VCR520 and CV965 were $3\frac{1}{2}$ inch magnetically focussed and deflected tubes with a screen of high purity potassium chloride. Their face was flat so that the image could be projected episcopically with $\times 8$ magnification by an f/1.5 lens on to the underside of the viewing table, using illumination from extended-arc mercury lamps with a total power input of about 1 kW. This necessitated forced air cooling of the tube face to keep its temperature around 16°C, which was found to be the optimum for displaying small signals with adequate contrast and persistence. The skiatron has the disadvantage that the trace made by a single transit of the beam fades very slowly. The persistence can be reduced by shutting off the cooling and so allowing the screen temperature to rise, and simultaneously subjecting the screen to low intensity electron bombardment. Even so, it required several minutes for 'permanent' echoes to be erased. Long term bombardment by the electron beam could cause permanent staining of the screen. For these reasons, use of the skiatron was limited in wartime to a relatively small number of 'constant heading' displays. It was not used in post war military or civil surveillance systems.

5.6 Postscript

Unlike microwave research and development, wartime activity on cathode ray tubes produced no dramatic change in technology. In the words of the senior engineers

'the demands on the cathode ray tube for radar applications are rather different from those imposed by television, so that much of the research effort was expended on projects which find little outlet in the latter field —— the very considerable work done on afterglow screen problems finds no use in television —— the great majority of the research work was directed towards electrostatic tubes, which in 1939 were almost entirely superseded by the magnetic variety for television purposes —— few improvements of significance have been obtained in magnetic tubes as a result of wartime research, and television tubes of the immediate postwar years are not likely to differ very radically from those made in 1939'.

With hindsight, it is not difficult to see how they came to underestimate the value of the advances in technology which resulted from their cooperative effort and that of all the other members of the CCRTD team. Quite apart from the innate caution of engineers, it would have been difficult to predict the rate of growth and the importance of peace time applications of cathode ray tube displays in industrial measurements, process control, and medicine.

The Plan Position Indicator proved to be one of the most important advances in military and civil radar technology.

Table 1   Basic Types of Cathode Ray Tube

* Types illustrated on following pages

Table 1   Basic Types of Cathode Ray Tube

| TYPE | | DATE in use | APPLICATION | SCREEN | | |
DATA	DISPLAYS			DIAMETER	COLOUR	PHOSPHOR
VCR84	CV1084	1938	Early CH	12″	Blue flash Yellow afterglow	Double layer (Zinc Sulphide:Silver exciting Zinc Cadmium Sulphide:Copper)
NC7	CV956	1939/40	Naval, Fire Control Radar Type 282 etc	12″	White, replaced by Green	Standard TV, replaced by Willemite (Zinc orthosilicate: Manganese)
VCR85	CV1085*	1939	CH	12″	Blue flash Yellow afterglow	Double layer
VCR131 (standard oscilloscope tube)	CV1131		CHL, Oboe	12″	Green	Willemite
VCR511	CV1511		CHL, GCI (Ground Controlled Interception)	12″	White, long afterglow	Calcium Phosphate: Dysprosium
VCR87	CV1087*		Mobile CH Range Amplitude Display (A-scan)	6″	Blue flash Yellow afterglow	Double layer
VCR97	CV1097*	1940	General Purpose, developed for ASV Range Amplitude Display	6″	Green	Willemite
VCR138A	CV1517	1941	AI MkIV (VCR97 with smaller bulb at screen end)	$3\frac{1}{2}$″	Green	Willemite
VCR514	CV1514	1941	AI MkVII (replaced by VCR138)	$3\frac{1}{2}$″	Green	Willemite
VCR521	CV1521		AI MkIX (VCR138A with afterglow screen)	$3\frac{1}{2}$″	Blue flash Yellow afterglow	Double layer
VCR524	CV1524		AI MkIX (successor to VCR521)	$3\frac{1}{2}$″	Blue flash Yellow afterglow	Double layer
VCR139A	CV964*	1942	General purpose, Rebecca, also used as low voltage tube in gliders, etc.	$2\frac{3}{4}$″	Orange	Zinc Sulphide: Manganese
VCR526	CV1526	1944	Range Amplitude Display, (daylight viewing). Rebecca Mk 3N	$2\frac{3}{4}$″	Green	Willemite
VCR527	CV1527	1942	Oboe monitor	$1\frac{1}{2}$″	Green	Willemite
VCR522	CV1522*	1942	General purpose (139A with smaller bulb) also used for projection on windscreen	$1\frac{1}{2}$″	Green	Willemite

PICTURE DISPLAYS

VCR140(A)	CV1140	1940	PPI 2–10 r.p.m.	12″	Blue flash Yellow afterglow	Double layer
VCR516	CV1516*	1941	PPI 2–10 r.p.m. (VCR140 with smaller bulb and screen)	9″	Blue flash Yellow afterglow	Double layer
VCR517	CV1517	1942	PPI 60 r.p.m. H_2S, ASV (modified VCR97)	6″	Orange/Yellow	AB mixture (Zinc Cadmium Sulphide with 12% Cadmium Sulphide: Copper + Silver)
VCR530	CV1530*	1944	PPI 60 r.p.m. H_2S, ASV (replacement for VCR517)	6″	Orange/Yellow	AB mixture
	CV254	1944	PPI 2–10 r.p.m. Naval radar (replacement for CV516)	9″	Blue flash Yellow afterglow	Double layer
VCR520	CV1520	1943	PPI 2–10 r.p.m. (Skiatron)	$3\frac{1}{2}$″	Dark trace	Potassium Chloride
NC17	CV965*	1943	PPI 2–10 r.p.m. (Skiatron)	$3\frac{1}{2}$″	Dark trace	Potassium Chloride

* Illustrated on following pages

CV 1097

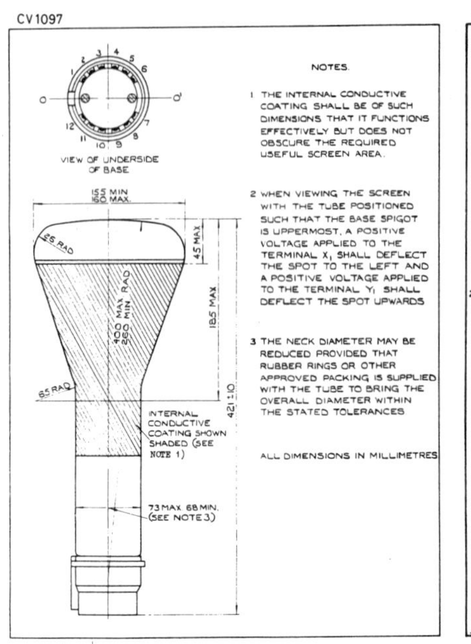

CV 964

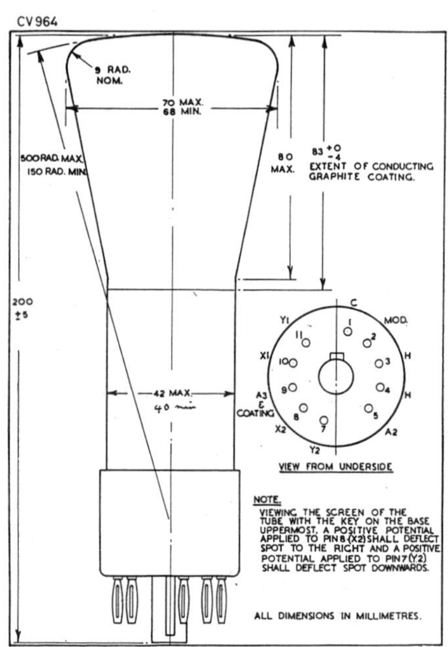

CV 1522

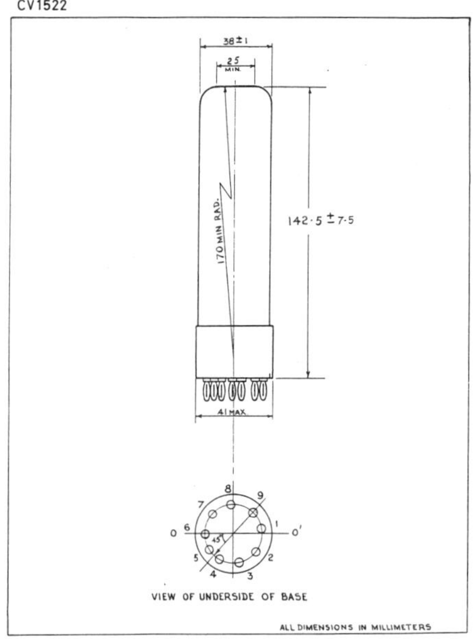

CV 1516

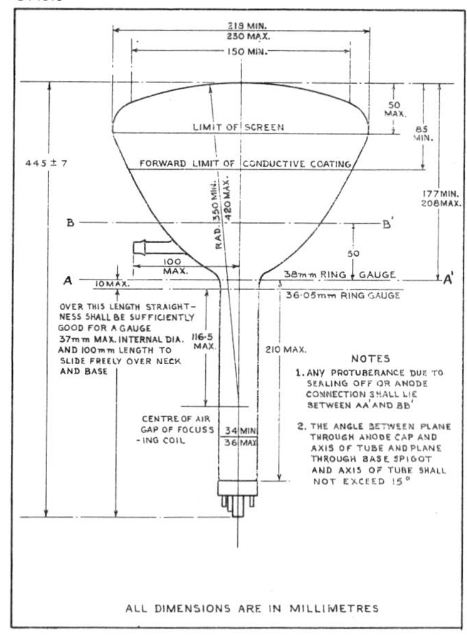

CV 1530

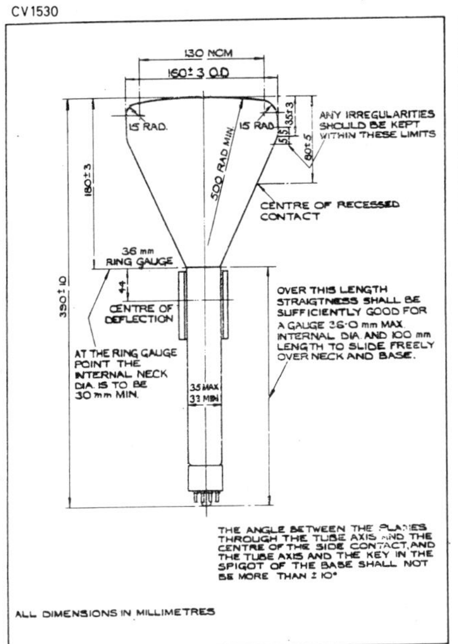

CV 965

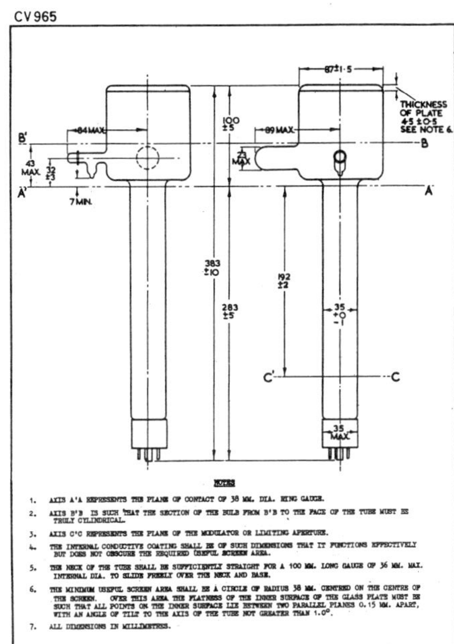

CV 1087

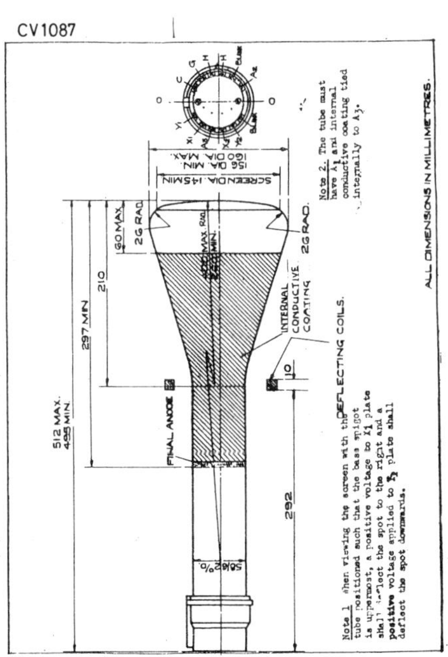

CV 1085

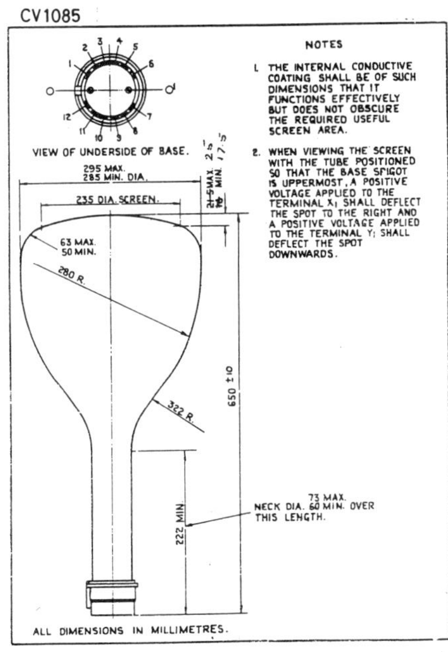

Chapter 6

Modulator Switches

6.1 Grid Modulation

Triodes and tetrodes are amplitude modulated devices. Those used in metric radar transmitters differ from their counterparts in communications equipment only in having proportionately larger cathodes to increase the available primary emission and so the peak power output, and being designed to withstand higher anode voltages. Thus metric radar transmitters can be turned on and off by a DC pulse applied to the grid of the oscillator, which is the first and sometimes the only valve in the RF power generating chain. During the interval between pulses, the anode current in all the valves in the chain is reduced to zero by a DC bias voltage on their grids. When amplifying or buffer stages are connected between the aerial system and the oscillator, they are turned on by the RF pulse which it generates.

In CH and other land based systems which operated in the range 20 to 50 MHz and generated up to 600 kW peak in 5 to 30 microsecond pulses with a repetition rate of 50 pps, the DC pulse was applied to the 'earthy' end of the RF circuit connected to the grids of the push-pull oscillators. [An alternative method of 'grid modulation' was used in the Army GL1 and GL2 transmitters, in which a positive pulse was applied via a capacitor and resistor in parallel to the grid circuits of the biassed-off oscillators, so raising their potential from negative to near zero. This initiated the RF pulse, which ceased when the grid current of the oscillators charged the capacitor to a negative voltage at which oscillation could no longer be maintained. During the pulse interval, the capacitor was discharged by the resistor while the anode current in the oscillators was cut off by the DC bias voltage. This technique, known as 'squegging', was used to generate peak powers up to 150 kW in the range 50 to 80 MHz and 80 kW at 200 MHz in pulses with durations between one and five microseconds and repetition rates up to 1000 pps.]

None of these 'grid control' systems required major development of modulator valves because DC pulses with the required characteristics could be generated by hard valves and mercury thyratrons used in communications and industrial equipment, or variants thereof.

6.2 Cathode Switching

In the Type 79 Naval radar, which operated at 43 MHz and generated 70 kW peak in 5 or 20 microsecond pulses with a repetition of 50 pps, the oscillators were switched on and off by a positive pulse applied to the grid of a similar but larger valve with its anode connected to their cathodes, its cathode connected to earth (the negative terminal of the high voltage power supply), and its grid to a negative supply which supplied the cut-off voltage. In parallel with the switching valves the cathodes of the push-pull oscillators were connected to earth by a high resistance and an RF bypass capacitor which provided a low impedance path for RF currents circulating in the oscillator assembly. After the pulse had passed, the cathode potential of the oscillators rose quickly to a value sufficient to suppress oscillation (their grids were connected to earth through their common RF circuitry). [Type 79 was the first radar system to use this method of modulation.]

The Naval Type 281 radar used the same modulation system as Type 79, except that four NT78 silica triodes connected in parallel as two pairs were used to modulate the oscillators, which generated up to one megawatt peak at 90 MHz in two microsecond pulses at a repetition rate of 50 pps. With 15 microsecond pulses, the power output was limited to 300 kW by the permissible maximum mean power input to the oscillator valves. Designed in 1939 by F.M. Foley of Signal School, NT78A was a longer version of NT57T, except that the tops of the twin filaments were rounded like those he used in NT86 to increase their effective area. Rated at 10 V 65 A, their available emission was about 50 A. Holding off 50 kV on the anode with a grid bias of -3 kV, this output current could be obtained by driving the grid to about $+1.5$ kV, which required an input of 5 A from the driving pulse generator.

> In the design of hard valve modulator switches (and in transmitting valves), a compromise has to be made between the grid bias voltage which cuts off the anode current at a specified anode voltage and the grid input power required to produce the desired anode current. The cut-off voltage is determined by the amplification factor μ, which is the ratio of the anode voltage and the minimum grid voltage at which no anode current flows, and represents the ability of the grid to limit penetration of the anode field into the grid-cathode space. Thus a high μ triode will require less cut-off bias than a low μ type. Because of this higher screening effect, the grid will have to be driven more positive than that in a low μ valve to obtain the same anode current. Thus the grid input power will be considerably higher. This requires the use of a more powerful driving circuit, and may cause problems with overheating of the grid and the consequent risk of uncontrolled grid emission. The low μ design has the disadvantage that it requires a higher bias voltage and a correspondingly larger driving voltage.
>
> If a tetrode is used in place of a triode, the screen grid G2, which operates at a fraction (typically about one tenth) of the anode voltage, limits penetration of the anode field into the region of the grid, so that less bias may be used to cut off the anode current. In the driven condition, the G2 voltage assists the flow of electrons through the grid, making the anode current almost independent of anode voltage. Thus the grid driving voltage and the input power can be much less than that for a triode operating with the same anode voltage. In effect the tetrode operates as two low μ triodes in series, combining a high cut-off capability with low grid driving power. To minimise loss of anode current caused by collection of electrons by G_2, the two grids must be accurately aligned.

6.3 Hard Valve Modulators

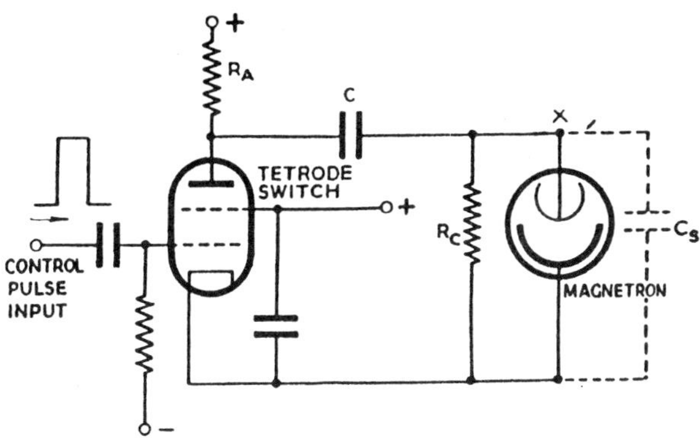

In a typical hard valve modulator, energy is stored during the pulse interval by charging the capacitor C through R_A and R_C while the tetrode is biassed beyond cut-off. The control pulse then drives its grid slightly positive, causing C to be discharged through the anode and magnetron impedances in series until the grid voltage is restored to the cut-off value. Tetrodes are preferred to triodes because their operating voltages can be chosen so that the variation of anode current with anode voltage is small enough for the fall of voltage across C to have no significant effect on the shape of the output pulse (C is only partly discharged by the output current).

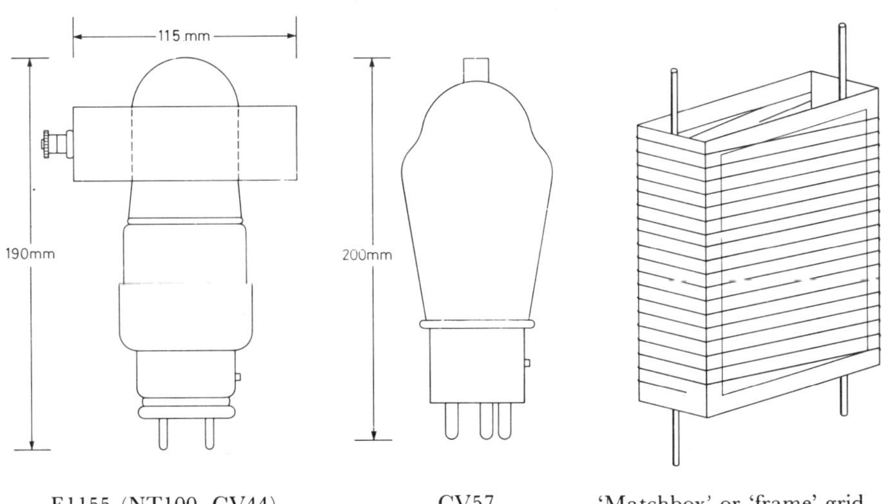

E1155 (NT100, CV44)   CV57   'Matchbox' or 'frame' grid

In 1940, when the E1189 and E1198 magnetrons became available for use in experimental radar systems, no line modulators (see 6.4) had been developed. Thus the first equipments used the GEC E1155 external anode tetrode to generate the magnetron input (10 A at 10 kV in microsecond pulses with a repetition rate around 1000 pps). Type Approved as NT100 in March 1941, E1155 was used in the Naval Type 271 S-band radar. In mid 1941, as CV44, it was flight tested in the early models of AI MkVII. For production reasons, it was replaced by CV57, an 'all-glass' envelope beam tetrode with a lower power rating (about 5 A at 10 kV). NT100 and CV57 used box section oxide cathodes, similar to but slightly larger than those in the KT66 beam tetrode (see 7.2). NT100 was fitted with four side by side, surrounded by a grid and screen grid of approximately rectangular cross section, spiral wound on support rods which passed through ceramic laminae at each end of the assembly. These also housed the ends of the cathode. CV57 used two cathodes side by side. The grids were wound on 'matchbox' frames and accurately aligned as in other beam tetrodes. A similar valve, the Mazda V1214 was developed by Ediswan at about the same time and could be used as an alternative to the GEC type. By the end of 1941, CV57 was Type Approved for use in AI MkVIII, two being used in parallel to modulate the CV64 magnetron. In mid 1942, it was superseded by the trigatron CV85 which was used in the Modulator64 to generate around 160 kW peak output (about one and a half times that of the hard valve modulator system). Modulator 64 was fitted to H_2S in October 1942.

The last and most efficient hard valve pulse modulator switch made at Signal School was the CV313 tetrode, designed by F.M. Foley in 1944. It used a filament and grid assembly similar to that in NT78A, fitted inside an anode of diameter 85 mm and length 190 mm. Unlike the earlier valves, the anode was formed from molybdenum sheet with four internal stiffening rings. The new construction increased the thermal radiation efficiency of the anode and also improved the thermal efficiency of the cathode. [The basket anode was highly transparent to radiation incident upon it at angles less than about 60°]. The G_1 and G_2 spirals were aligned to minimise loss of anode current and simplify cooling of the screen grid. With 50 kV anode voltage and 5 kV on the screen grid, the anode current could be cut off completely by a grid bias voltage between -2 and -3 kV. At zero grid potential the anode current was approximately 40 A. [Compare with NT78A above]. Intended originally for use in a magnetron test set, it was almost immediately adopted for the development model of the Type 992 target indicating radar, where two were used in parallel to provide the mean power required by the magnetron transmitter, which generated 2 megawatts peak power in 2 microsecond pulses with a recurrence of 1500 p.p.s. Development was not completed until 1946. After a considcerable number of years in service, CV313 was replaced by a line modulator using a hydrogen thyratron switch.

There was no other British development of hard valve modulator switches until requirements arose for replacements for American types used in commercial marine and aircraft radar.

6.4 The Line Modulator

Grid modulation, cathode switching and squegging all require that during the pulse intervals the RF power valves withstand without catastrophic flashover the high operating voltage applied to their anodes. [Many of the equipments incorporated spark gaps and other devices to limit these effects]. This immunity becomes progressively more difficult to achieve as the clearances between the electrodes are reduced to obtain efficient operation at higher and higher frequencies. Ultimately, it requires that the operating anode voltage be applied only for the duration of the pulse. Thus the triodes in high power transmitters operating at and above 600 MHz (50 cm wavelength or shorter) must be 'anode modulated'. For obvious reasons a similar technique has to be used with the magnetrons in microwave transmitters (except that the pulse is applied to the cathode because the anode is part of the waveguide assembly, which must be earthed).

Magnetrons pass very little current until the applied voltage reaches the threshold of oscillation. They then require that the operating current be kept as constant as possible to minimise 'frequency pushing' (the variation of output frequency with anode current). Thus their DC input pulse must have a rapid rate of rise (to permit accurate synchronisation of the leading edge of the output pulse with the time base circuits in the radar display), and a flat top. This was a major factor in stimulating development of a new modulation technology, in which an 'artificial line' or pulse network is charged to the required voltage and then discharged into the transmitter by a switch which ceases to be conductive before the next charging cycle begins (in some systems an auxiliary 'hold- off' switch in the charging circuit prevents premature discharge of the line). The pulse may be transferred directly into the load, or through a step-up pulse transformer when the required output voltage causes the line voltage to exceed the hold-off capability of the switch. [The line voltage is approximately twice that delivered to the load]. Alternatively, two pulse forming networks may be charged in parallel and discharged in series by a pair of triggered switches. (see Appendix 5 'Line Modulators').

Generation of pulses by the discharge of artificial lines was invented by O.L. Ratsey of Signal School in mid-1940, and used to drive the 'cathode switching' modulator valves in the Naval Type 281 radar system. He also designed an experimental line modulator using a length of high voltage coaxial cable as the energy store. Major contributions to development of the new technology were made by A.D. Blumlein, who invented the 'Blumlein' modulator in which two parallel charged pulse-forming networks are discharged in series by a single switch which is required to hold off only the voltage across one of them. In November 1940, K.J.R. Wilkinson of BTH was the first to use the parallel charge/series discharge technique, and in September 1941 devised the system of 'alternator charging' in which no rectifier is necessary.

A twin-line alternator charged modulator using two CV22 mercury thyratrons to generate 40 A at 25 kV in one microsecond pulses with a repetition rate of 400 pps was in production at the end of 1941 for the Army GL3 S-band radar system. About a year later it was replaced by a Blumlein modulator using an open-to-air triggered spark gap switch developed from a unit researched by MetVick in 1941. [Unlike the thyratrons, the spark gap switch required no 'warm up' time, and so permitted rapid start up of the equipment from cold].

6.5 Thyratron Switches

6.5.1 Foreword

A thyratron is a gas filled device in which no anode current flows until a positive voltage is applied to the negatively biassed grid*. Thereafter, the grid has no control over the rate of rise of the anode current, which is determined by the build up of a glow discharge in the gas, first in the grid-cathode space and then between the grid and anode. This is accompanied by a fall in the anode voltage to the DC equilibrium or 'maintaining' voltage, which for mercury vapour thyratrons is in the range 10 to 20 V. The anode current is limited only by the total resistamce of the external anode circuit, including the internal impedance of the power supply, and continues until the glow discharge is extinguished by reversal of the anode voltage.

> *If the anode voltage is raised above the 'hold off' level before the positive voltage is applied, there will be an uncontrolled flashover between the anode and grid and hence to the cathode, with a high probability of catastrophic damage to the tube.

6.5.2 Mercury Thyratrons

Because they could pass high current (up to 100 A) with a low voltage drop across the tube, pairs of mercury thyratrons were used extensively pre-war as grid controlled full wave rectifiers providing DC power for welding gear and electric motors. [Many modern equipments use solid state devices for the same purpose]. By varying the time of application of the positive voltages to the grids, and so the fraction of the half cycles of the AC input during which each tube was turned on, the output voltage or current could be varied and accurately controlled from zero to the available maximum without the waste of power which would result from the use of a variable resistor to achieve the same effect.

When a mercury thyratron is used in a line modulator (see Appendix 5) to generate high voltage pulses with a duration measured in microseconds, the anode voltage falls continuously, but remains above the equilibrium level, reaching a minimum between 60 and 100 V before the discharge is extinguished by a negative overswing produced at the end of the pulse by a deliberate mismatch between the load and the line. Some of the anode current is carried by electrons moving from the cathode to the anode, some by electrons leaving the surfaces of the grid as others are collected by it. Thus it is difficult to describe with certainty the mechanism of build up and decay of the anode current.

In a typical mercury vapour thyratron switch such as CV22, the graphite anode is completely shielded from the cathode by the grid cylinder assembly, which contains the glow discharge and so protects the glass envelope from damage by electron or ion bombardment, and also acts as a secondary heat shield for the cathode. To prevent uncontrolled discharges passing outside the grid cylinder from anode to cathode, the top of the cylinder is capped by a welded on disc clamped round the glass sleeve surrounding the anode stem. [This also provides mechanical support for the whole assembly]. A perforated graphite disc (sometimes known as a 'pepper-pot' grid) provides a path for electrons moving towards the anode under the influence of its electric field, and for ions diffusing in both directions.

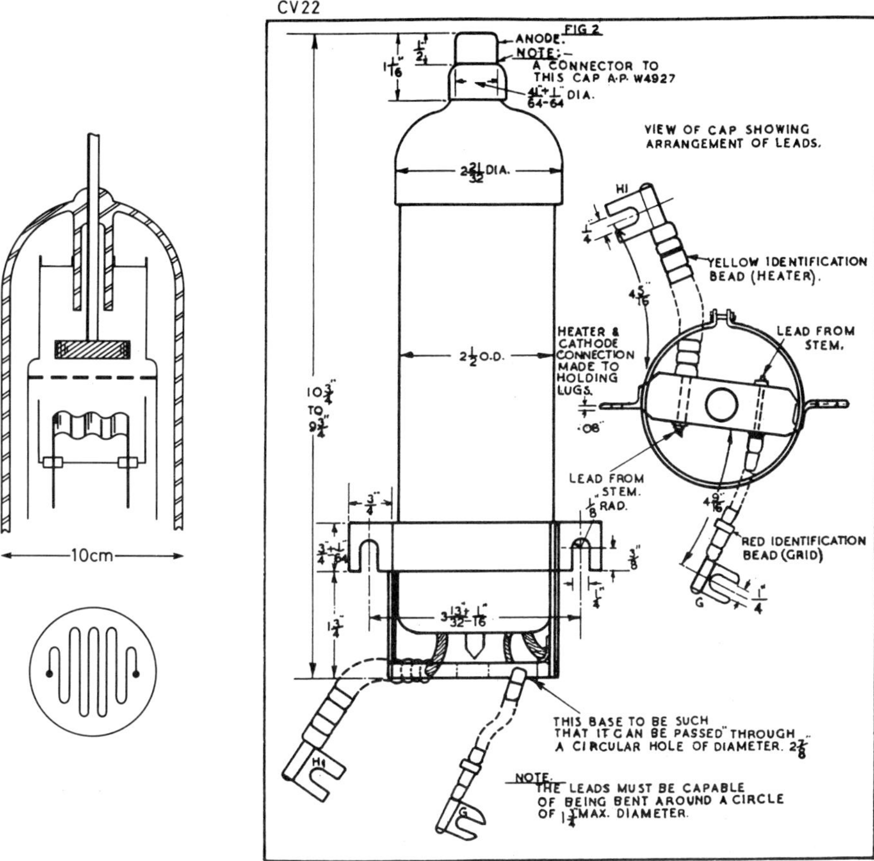

When a positive voltage is applied to the negatively biassed grid, electrons
move towards it, making ionising collisions with the mercury atoms so that a
plasma (a mixture of electrons and ions) fills the space below the grid disc. (This
manifests itself as a glow discharge in the grid-cathode space). Because of the high
mass and low mobility of the ions, this may take more than one microsecond.
When the plasma is established, electrons are drawn through the grid by the
anode field, to be followed by ions diffusing into the grid-anode space to neu-
tralise the 'space charge' effect. This causes a rapid build up of the anode current
to its maximum value. [It is recorded that in a typical case (CV22), anode current
begins to flow 1.5 microseconds after the grid begins to draw current from the
primary source, and then rises to its maximum in a much shorter time, estimated
as being in the range 0.05 to 0.1 microsecond. The rise time appeared to be almost
independent of this maximum].

The 'hang fire' effect is one reason why it is difficult to obtain and maintain
the accurate synchronisation required for generation of high voltage pulses by
pairs of mercury thyratrons in parallel charge – series discharge line modulator

systems. (see Appendix 5) The rise time of the anode current precludes their use in systems requiring a pulse duration below one microsecond.

During the interval between one pulse and the next, the plasma must decay so that the rise in network voltage does not induce premature firing. This requires that sufficient time be allowed for ions to diffuse to the surfaces of the grid assembly and there recombine with free electrons. Because of the low mobility of the ions, this 'recovery time' is typically greater than one millisecond, so that mercury thyratrons were used only in systems with pulse recurrence frequencies no greater than 500 pps.

Because of the high voltage across the tube during the early part of the discharge, ions at the edge of the plasma facing the cathode will move towards it with energies high enough to damage its coating and so shorten its life. This effect can be minimised by arranging that the grid-cathode spacing is as small as practicable (to reduce the voltage drop), that the cathode has the maximum possible area, and that the electron emission is 'space charge limited', ie that there is a 'sheath' or high concentration of electrons surrounding it, from which only a fraction is required to carry the anode current. It was observed that the cathode surface in modulator switches could be contaminated with nickel sputtered on to it by electron bombardment of the grid assembly during the 'priming' phase of the discharge, and that mercury entered the oxide coating. Thus one important aspect of the development of mercury thyratrons for pulse applications was the incorporation of cathodes with a much larger effective area than those in the industrial tubes from which their design was derived.

The rate of build up and decay of the plasma is determined by the vapour pressure. If it is too low, the rate of rise of the anode current is reduced, and the risk of damage to the cathode increased because ions with high energy will be moving towards and possibly bombarding the cathode for a longer time at the beginning of the glow discharge. If it is too high, the recovery time will increase because of the reduction in mobility and mean free path of the ions in the plasma. There will also be a reduction in the hold-off voltage because the mean free path of electrons released as field emission from the grid may be short enough for ionising collisions to occur before they are collected by the anode. Thus the vapour pressure must be kept within close limits to obtain and maintain the required combination of characteristics. For this purpose, the tubes contained a small amount of liquid mercury, the vapour pressure being controlled by maintaining the whole or part of the envelope at a chosen 'condensation temperature', typically around 60°C. This required that the tubes be housed in a temperature controlled enclosure or that a jet of air controlled by a contact thermocouple attached to the glass be directed at the envelope to produce a local 'cold spot'. In either system, the time required for the vapour pressure to stabilise at the operating level was between two and three minutes. This limited the use of mercury thyratrons to Naval and land-based mobile radar systems where the equipment could 'stand by' with the thyratrons kept at their operating temperature with the cathode heating power switched on.

The first equipment to incorporate a line modulator was the Naval Type 281 90 MHz radar. Designed by O.L. Ratsey in mid-1940, the modulator provided the grid driving pulse for the paralleled NT78A triodes which controlled the

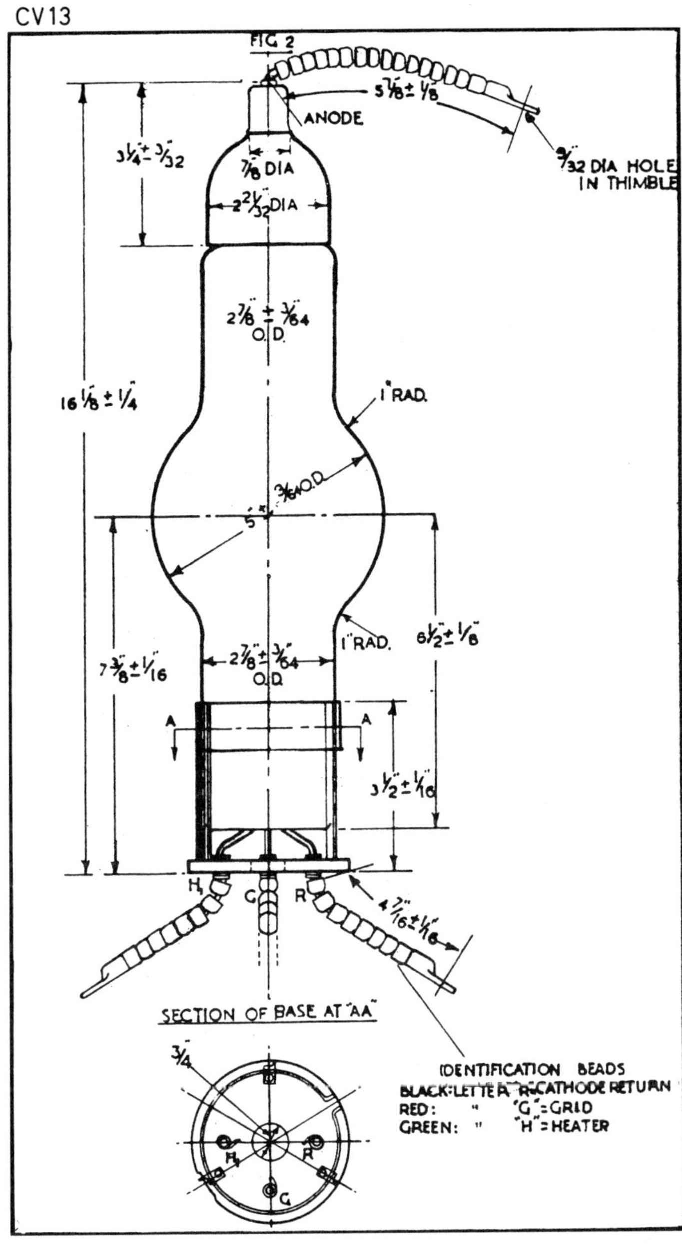

transmitting tubes. The switch was a modified BTH BT9 industrial mercury thyratron, later known as CV13.

Thereafter, development of pulse generator tubes for Service applications by the groups led by H.deB. Knight at BTH and H.G. Ramsay at GEC was

CV 12

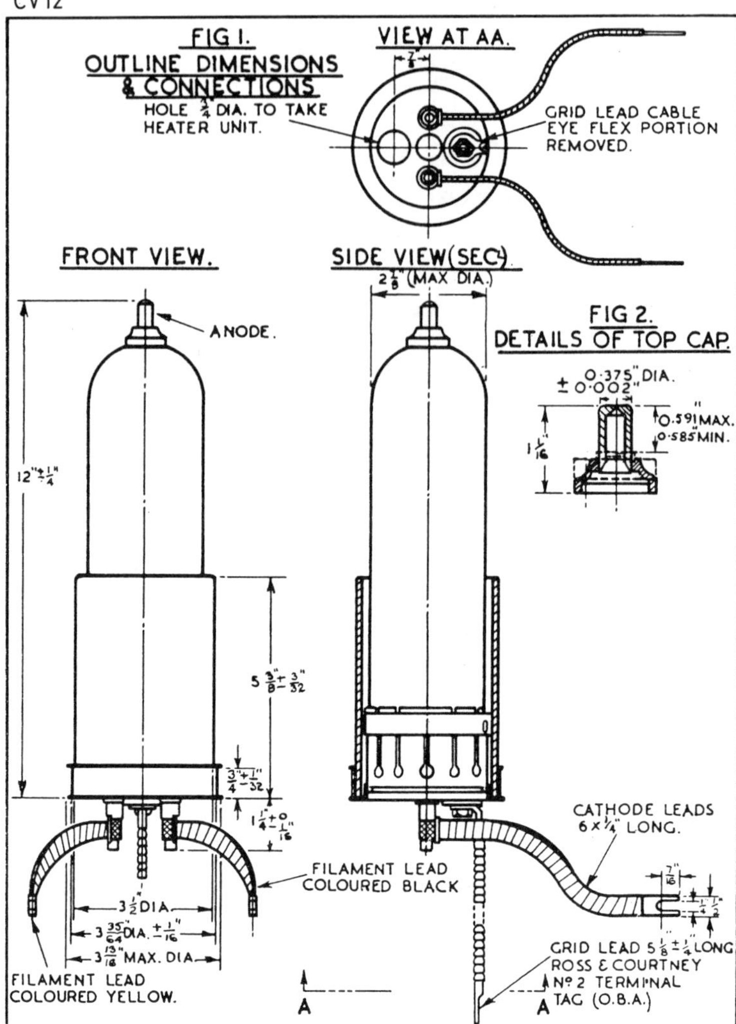

concentrated on improving mechanical strength, increasing the hold-off voltage and peak current rating, and extending the operating life, which was limited by cathode failure. The teams worked independently but were closely in contact, so that if necessary a tube designed by one group could be used as a replacement for one made by the other. [One example of this was CV12 developed and manufactured initially by GEC. When a second source was required, BTH made a 'plug in' replacement derived from their BT49.]

Because they are turned on for periods not greater than 20 microseconds at intervals typically not less than 2 milliseconds, the mean power rating of pulsed thyratrons is a small fraction of that applying to their near equivalents designed for industrial use. This is offset by the increase in cathode heating power arising from the incorporation of larger cathodes. Even so, the total input power of the pulsed tubes was less than half that of industrial types, with a corresponding easement of cooling requirements. Thus the tubes could be built in cylindrical envelopes not much greater in diameter than the grid cylinders. This in turn simplified equipment design and made possible improvements in the mechanical strength of the tubes themselves.

The first thyratrons used in line modulators (BT9 and BT9B) had indirectly heated 'pot' cathodes with an internal oxide coating. These had the disadvantage that the plasma tended to penetrate only a short distance into the pot, which reduced the effective area of the cathode. To overcome this, the BT45 and BT49 switches developed by BTH used directly heated folded or spiral nickel ribbon cathodes, with the oxide coating applied to nickel gauze welded on both sides of the strip. This resulted in a large increase in the emitting area of the cathode. The geometry of the assembly facilitated build up of ionisation in the grid-cathode interspace.

Because of the fundamental limitations on their characteristics, little more could be done to improve the performance and widen the application of mercury thyratrons in line modulators. They were not used in peace time military or civil radar systems.

Mercury Thyratrons – Characteristics and Applications

Type No	Prototype	Rating	Equipment	used with Valve	Date in Service
CV12	GEC E1191 BTH BT49	15 kV 200 A (second source)	Type 277	CV76	1943
CV13	BTH BT9B	16 kV 120 A			
CV1145	BTH BT9A	18 kV 40 A	Type 281	NT78A (modulator switch)	1941
CV22	BTH BT45	20 kV 50 A	Type 291 Type 271Q GL3 (two CV22s)	NT99 CV56 CV120	1941 1942 1941

6.5.3 Hydrogen Thyratrons

Because of their much lower mass (one hundred times less than mercury) hydrogen ions have corresponding higher mobility, and so provide the means for obtaining a very rapid rate of rise of the anode current in pulsed modulator switches (up to 5000 A per microsecond). The higher mobility also reduces recovery time commensurately. Thus the use of hydrogen in place of mercury in line modulator switches makes possible the design of systems operating with pulse durations below 0.1 microsecond and recurrence rates measured in thousands of pulses per second. [This advantage is offset to some extent by the higher voltage drop across a hydrogen tube (the ionisation potential of hydrogen is higher than that of mercury), but the effects of ion bombardment are reduced because of their lower mass.]

For these reasons, attempts were made in late 1942 by BTH and GEC to develop hydrogen thyratrons as replacements for their established mercury switches. These were unsuccessful, largely because the hydrogen filling was taken up by the electrode structures within a short operating time.

About two years later, this problem was solved by Egerton and Germeshausen of MIT, who used a titanium hydride capsule as the reservoir for hydrogen. Described as a 'molecular sieve', titanium hydride has the property that hydrogen atoms are retained within the crystal lattice so that equilibrium is established with the external gas pressure at any given temperature.

A small capsule can retain large quantities of 'hydrogen gas', releasing it to maintain the equilibrium as the gas is removed during operation of the tube. Thus the operating pressure can be held within the required close limits by control of the capsule temperature.

Postscript

By 1946, mercury line modulator switches were obsolete. BTH and GEC were developing hydrogen tubes based on American technology.

6.6 Spark Gaps and Trigatrons

6.6.1 Rotary Gaps

Voltage breakdown between a pair of electrodes is the simplest means for discharging a pulse forming network into a load. Operating in air with cold electrodes, it requires no warm up time and no system for accurate control of the vapour pressure during its operating life. Because the mobility of ionised air molecules is more than ten times higher than that of mercury, the spark gap switch permits operation with shorter pulses and higher recurrence frequencies.

The discharge can be initiated either by movement of one electrode towards the other, or by an auxiliary discharge between one electrode and a trigger mounted within it and so protected from the high current pulse.

Rotary spark gap switches were used in experimental line modulators from early 1940. They comprised a pair of electrodes, one on each side of a rotating bakelite disc which carried a number of pins which caused breakdown as they passed between the fixed electrodes. [Some systems had several sets of fixed electrodes operating in series to increase the operating voltage without risk of premature flashover as the pins approached the gaps.] Using these techniques, it was possible to obtain power outputs up to a few megawatts at 40 kV. Only one system of this kind was used operationally. Developed in 1942 by T.S. England of TRE, the RAF AMES11 ground based 50 cm radar transmitter used a rotary gap switch in which the disc carried six pins and rotated at approximately 5,000 rpm. The power output was 250 kW peak at 7 kV in 2 microsecond pulses with a recurrence around 500 pps. By adjusting the spacing of the fixed electrodes as the gap was increased by spark erosion of their tungsten tips and the tungsten pins, the equipment could be operated for several thousand hours between overhauls. Rotary spark gap switches have the inherent disadvantage that 'jitter' or random variation of the pulse interval may amount to 100 microseconds. For this reason, AMES11 included means for changing over to a 'trigatron' switch when accurate range measurement was required.

6.6.2 Triggered Spark Gaps

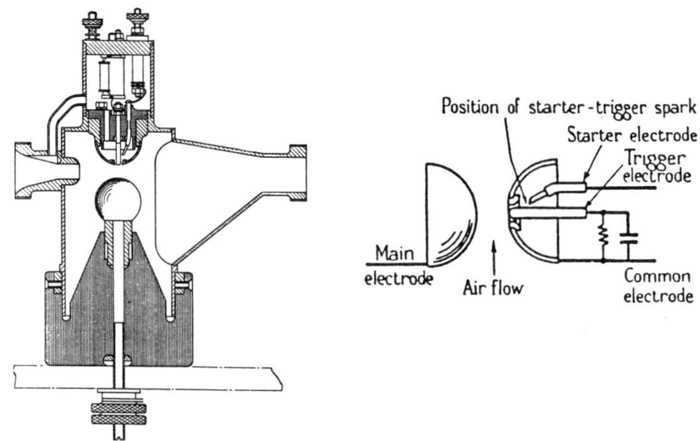

If, as in thyratrons, the discharge is initiated by a trigger pulse, spark gap switches can be used with advantage as replacements for mercury tubes in high power line modulators. [The spark gap retains its 'instant' start up and the ability to generate shorter pulses with higher recurrence rates].

In the four-electrode unit developed by MetVick in 1941, the main discharge is initiated by breakdown between the trigger electrode and the anode, which follows the application of a positive pulse to the starter electrode. (the RC circuit causes the priming discharge to be extinguished a short time later). Air is blown through the gap to cool the electrodes and scavenge the remnants of the glow discharge, including the products of erosion of the cathode.

In 1942, a Blumlein modulator (see Appendix 5) with a spark gap switch was fitted to the Army GL MkIII S-band radar. Generating a peak output of 40 A at 25 kV to drive the CV76 magnetron, it replaced a twin line system using a pair of CV22 thyratrons. The change was made to overcome problems with short life caused by over-volting of the mercury tubes, and to provide rapid 'start up' without keeping the modulator on 'stand by'.

Similar systems were fitted to the Naval Type 274 and 275 S-band gunnery radar systems, which were in operational use from mid-1943. The modulator provided a peak power output of 120 A at 9 kV in half microsecond pulses, which was fed to the magnetron through a 3:1 step-up pulse transformer.

6.6.3 Trigatrons

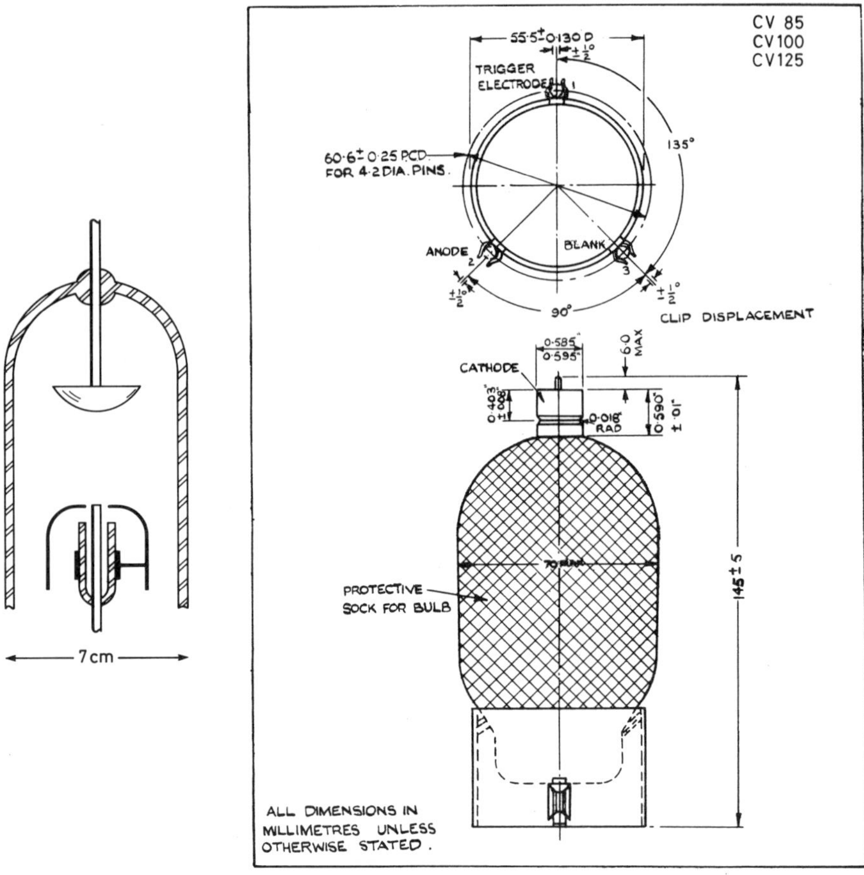

Trigatron

Because of the need to operate at high altitudes, spark gaps operating in air could not be used in airborne radar. [Open-to-air spark gaps were also vulnerable to corrosion of the metal parts and degradation of the insulation by ozone and nitrogen oxides formed during the discharge, particularly in the presence of water vapour].

Because it made possible the design of airborne transmitters using strapped magnetrons to give output powers limited only by that available from the aircraft generating system, development of the 'Trigatron' by J.D. Craggs, M.E. Haine and J.M. Meek of MetVick (Metropolitan Vickers Electrical Co.) was a major contribution to mobile radar technology. The MetVick team had previously studied high power three-electrode open-to-air units with tungsten triggers and molybdenum electrodes carrying the pulse output current.

Early in 1941, experiments with molybdenum electrodes in hydrogen showed that lives of about 100 hours could be obtained from units giving 300 kW peak output in one microsecond pulses at 500 pps. Life was terminated by the build up of molybdenum 'spikes' which could bridge the trigger gap. There were also variable 'hang fire' effects and rapid wear of all the electrodes.

Following a report by S.C. Curran of TRE that an experimental demountable tube filled with argon showed much less wear of the electrodes, attention was concentrated on sealed off systems filled with argon at pressures of a few atmospheres. These operated successfully for about fifty hours, but persistence of ionisation limited the pulse recurrence frequency unacceptably. It was known that oxygen can act as a quenching agent for metastable argon ions, and so experiments were continued with argon at pressures up to four atmospheres containing oxygen at partial pressures up to 20 mm Hg.

By January 1942, a satisfactory sealed-off design was available for manufacture. To ensure that the trigger discharge took place at the top of the electrode, it was enclosed in a pyrex tube sealed to its support at its end remote from the opening in the positive electrode. To safeguard operators from injury arising from breakage of the glass envelope, this was covered by a 'sock' of mesh fabric varnished to the glass. Lives in excess of five hundred hours were obtained with ouputs up to 600 kW peak in one microsecond pulses with recurrence rates up to 2000 pps. This was acceptable operationally. The table below shows the important characteristics and applications of the three types manufactured in quantity by the Cosmos Manufacturing Company Ltd (an associate of MetVick, who developed and manufactured the 'Mazda' range of receiving and other low power tubes used in broadcasting and communications equipment, including the high gain pentodes used in the IF strips of early radar receivers).

Table   Trigatrons — Characteristics and Applications

Type	Peak output kw	Pulse length μ sec	Line voltage kV	Pulses per second	Application
CV85	170	1	7.2	1200	H_2S, AI
CV100	250	2	16	400	AMES 11
CV125	500	1	12	800	H_2S, ASV

In mid-1942, CV85 was Type Approved for use in Modulator 64. Generating 160 kW peak output it was used to drive the CV64 magnetron in H_2S. Early in 1943, it was followed by CV 125 in Modulator 158, which provided outputs up to 500 kW. In total more than 100,000 were manufactured.

Like mercury thyratrons, trigatrons were superseded by hydrogen thyratrons, which required less trigger power and gave longer lives. They found no application in peace time.

Chapter 7

General Purpose Valves

7.1 Foreword

In addition to the active devices developed specifically for operation at frequencies above 50 MHz, or to generate high powers in transmitters and pulse modulators, all wartime radars used large numbers of 'general purpose' valves, for example in RF, IF and video signal amplifiers, in time base circuits, and in the low power driving stages of high power pulse generators. With few exceptions, these general purpose valves were established indirectly heated commercial types developed by British valve makers in the 1930s, and manufactured in quantity for use in radio and television receivers and telecommunications equipment. (Some rectifiers used directly heated tape filament cathodes.)

7.2 Early Days

In radar systems such as Chain Home, Type 79, and GL1, the general purpose valves were developed in the early part of the decade by Cossor, Ediswan, (the Edison Swan Manufacturing Company, also known as 'Cosmos', who used the brand name Mazda), and MOV (the Marconi Osram Valve Company), then jointly owned by GEC and the Marconiphone Company. They were physically large (up to 16 cm high overall including the pins and between 5 and 6.5 cm diameter), and fitted with 4-, 5-, or 7-pin moulded bakelite bases cemented to the lower end of the glass envelopes, which were tapered to fit. The pins themselves were about 2 cm long, and radially compressible for about half their length, so providing a sprung contact with the sockets in the valve holder, through which connections were made to the external circuitry. Partly because of their size and shape, which made the use of screening cans uneconomic of space in equipment, RF pentodes and high gain triode amplifiers were screened by hot spraying zinc particles onto the outside of the bulb, covering in so doing a thin metal disc connected to one of the pins. This process, and that used for attachment of the base, was satisfactory when the valves were operated in a temperate environment, but could not resist the hot and humid conditions within some military equipment – the metallising flaked off and could cause short-circuits when the units were subject to shock and vibration; bulbs could be detached from their bases when attempts were made to remove valves from their sockets, particularly after they had been in use for some time.

At that time, there was no international agreement on outlines, pin configuration and heater voltage. European and British valve manufacturers used similar bulb and base assemblies but different pin configurations. American 'vacuum tubes' were smaller, up to 14 cm overall and 5.5 cm diameter, with 4, 5, 6 or 7 rigid pins in moulded bases, again with their own configurations. The choice of heater voltages was arbitrary, based on the choice of cathode dimensions and the availablity of technology for fabricating hum-free heaters for AC operation (to minimise modulation of the anode current by their magnetic field), and for coating them with an adherent layer of ceramic insulation. British manufacturers used 4 V 1 A heaters for signal amplifiers and 4 V 2 A for power valves with outputs up to about 8 W. American tubes, which had smaller electrode systems, operated at 2.5 V, 1 or 2 A.

7.3 The Beginning of Standardisation

In 1934, RCA, which was one of the largest American manufacturers of electronic tubes and equipment, introduced a range of 'all metal' receiving tubes. These were up to 11 cm long overall, and used cylindrical steel envelopes with diameters 3.4 cm for the lower power types and 4.5 cm for power tubes such as the beam tetrode 6L6. Up to 8 connecting leads from the electrode assembly were sealed through a ring of glass beads in nickel iron eyelets welded to the steel base of the tube, and soldered to rigid pins moulded into a bakelite disc, which was attached to the envelope by 'crimping' the skirt of the steel base into recesses in the periphery of the moulding. The disc also had a central spigot with a narrow vertical fin which aligned the tube with its socket before the pins entered the holes housing the spring contacts which connected them with the external circuitry. (The spigot protected the exhaust tube. The socket had a central 'keyhole' for the spigot). When a top connection to the grid of RF amplifiers was required, this was made through a top cap mounted on a bakelite disc above a bead assembly welded to the top of the envelope. The first tubes in the range were designed with 6.3 V heaters so that they could be operated from AC mains or from motor vehicle battery supplies either 'on charge' or free standing.

Because the new type had smaller electrode systems and closer grid wire to grid wire and grid to cathode spacing, the signal amplifiers provided higher gain than their predecessors. Having shorter connections between the electrodes and the external circuitry, they operated efficiently at higher frequencies. The steel envelope was an effective screen. It also made it easier to cool the anode of power amplifiers such as the beam tetrodes 6V6 and particularly 6L6, which surpassed its competitors in power output and gain. [The grids in beam tetrodes are aligned to minimise interception of electrons by the outer grid or 'screen'. The purpose is to reduce the power dissipated by the screen wires and their support rods, and so make it easier to cool them by 'flags' spot-welded to the rods above the top mica. The geometry of the beam forming plates and choice of the separation between the screen (G_2) and the anode gives the beam tetrode characteristics similar to those of a pentode without the need for a third grid or suppressor, which in pentodes prevents the interchange of secondary electrons between the screen and the anode as the anode voltage falls below and then rises above that of the screen when an alternating voltage is applied to the grid (G_1)]

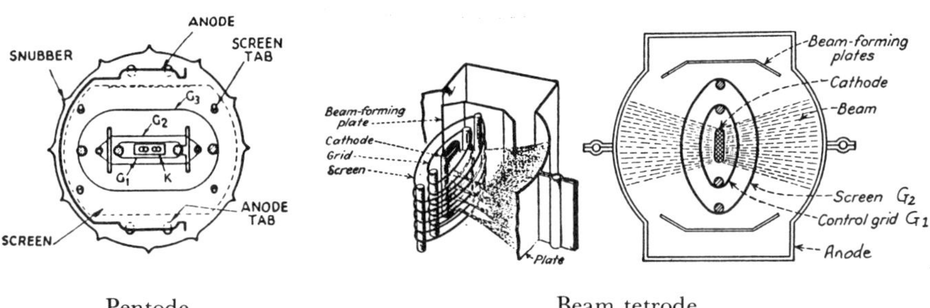

Pentode   Beam tetrode

Following accepted American practice, the characteristics of tubes in the new 'Octal' range were adopted by other US manufacturers. Some copied the RCA design under licence. Sylvania developed all-glass equivalents. [At about the same time, Philips and Telefunken in Europe each introduced a range of glass valves considerably smaller than their predecessors. Following European tradition, they were not interchangeable. The Philips types had a ring of vertical strip contacts on the cylindrical surface of the base. Telefunken used short radial pins near its foot.

Production of the 'all metal' tube required a massive investment in new manufacturing technology and equipment, particularly for making the 'ring welds' which were fundamental to the assembly process. In comparison with their glass equivalents, the metal tubes had inherent disadvantages, such as the larger number of seals, the difficulty of making them vacuum tight in large scale production, and the risk of porosity of the steel envelope, which had to be covered by a layer of paint which could prevent corrosion and resist abrasion. Glass is a far better material for fabricating vacuum tight envelopes. [Later RCA designs replaced the bead seals in the base by a glass 'button' with a ring of leads passing through it]. No European valve manufacturer followed RCA.

Before the end of 1937, Philips had fitted Octal bases to valves in their side contact range, so making them compatible with American types, either as plug-in replacements or functional equivalents. MOV had designed a compatible range as an alternative to their 4 V series. Thus the 'Octal' format became the first international standard for receiving tubes. For reasons which now seem difficult to understand, the new Mazda range was designed with 4 V heaters and 'British Octal' bases, so that they were not interchangeable with comparable types in the international series.

By 1938, European and British manufacturers, including Mullard who were closely associated with Philips, were developing and manufacturing large numbers of octal tubes. Not all of these were plug-in replacements for American types; some had superior performance.

From late 1938, because of its high performance, and despite its idiosynchratic format, the Mazda SP41 was one of the most commonly used general purpose valves in British radar and communications equipment, operating equally well as a low noise high gain RF amplifier, as an audio or pulse signal amplifier, and as a waveform generator in display time base circuits. Some months after the outbreak of war, in response to pressure from designers of military equipment, it was replaced by VR65A, which differed from it in having a 6.3V heater and an International Octal base. The KT66 power amplifier tetrode, developed by MOV in 1937 as a competitor to the American 6L6, was also widely used. A variant was fitted with a top cap anode connection to increase its voltage rating. Typical applications were as the driver for the deflection systems in data display tubes and the deflector coils of PPIs, in the output stages of audio and pulse amplifiers and low power servo systems, and as a driving stage in pulse modulators. Wherever possible, octal valves from MOV and Mullard were used interchangeably, so providing a safeguard against loss of production by either manufacturer.

7.4 A New Technique

The most important advance in general purpose valves in the late 1930s was the introduction of the 'Loctal' EF50 signal amplifier pentode by Philips. [Manufactured in Britain by Mullard and used by Pye, the EF50 made possible the design of a 'straight' television receiver, where a series of bandpass amplifiers provided the required sensitivity and frequency response. Other British manufacturers used a superheterodyne system to receive the 45 MHz transmissions.]

Up to that time, with the exception of all-metal types, receiving valves and small power amplifiers used an assembly technique in which the connections to the heater, cathode, grids and anode were made by wires sealed through a flat 'pinch' at the top of a short length of glass tubing (the 'stem'). The other end of the stem was sealed into the open end of a glass bulb which formed the major part of the vacuum tight envelope. [The tube connecting the valve to the pump was sealed into the stem below the pinch.] The electrodes were located and supported by support wires passing through mica discs at each end of the assembly. The cathode with its internal heater was anchored in the lower mica and free to slide in the upper. To provide the necessary screening between the grid and the anode of SP41 and other RF signal amplifier pentodes, the top mica carried a metal disc connected to the outermost grid and through it to the cathode, and the grid connection made by a wire sealed through a bead at the top of the envelope as the first stage in the sealing and exhaust process. The upper mica disc had a number of snubbers (triangular projections round its periphery) so that the glass envelope provided mechanical support for the electrode assembly when it was pushed into the upper part of the bulb. After the exhaust process was completed, the tube was sealed off by melting and severing the exhaust tube. The grid connection was then soldered through a top cap cemented over the bead, and the remaining leads through the pins of an 'Octal' base cemented to the lower end of the envelope.

This form of construction had two major disadvantages:- the length of the leads from the electrodes through the base and its associated socket to the external

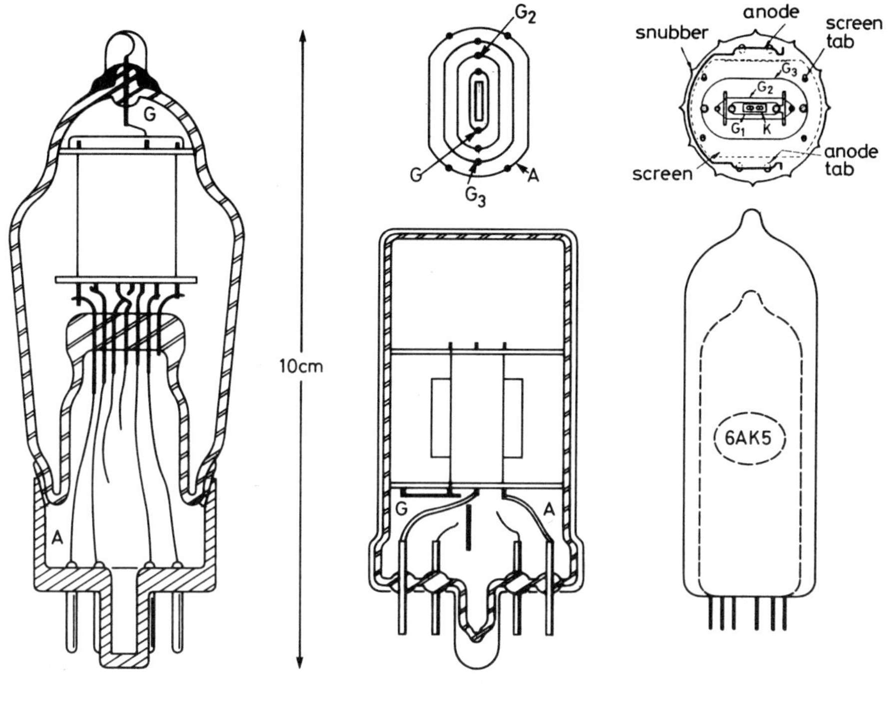

Octal SP41   Loctal EF50   7 and 9 pin miniatures

circuitry, and the use of a top cap connection to the grid, which required that a
screened flexible lead be used to make the connection. Both of these limited the
upper frequency of operation to about 30 MHz.

The electrode assembly of the EF50 signal amplifier and other valves in the
'Loctal' series was a single-ended structure mounted close to a ring of nine pins
in the moulded glass base. To reduce the inductance of the leads connecting the
electrodes to the pins, these were fabricated from narrow strips of nickel. Efficient
screening of the grid from the anode was obtained by arranging that their
connections were on opposite sides of the ring and by attaching a metal shield to
the cathode and suppressor (G_3) connections. As in the 'Octal' series, the electrodes were located and supported by mica discs at each end of the assembly. The
exhaust tube was sealed through the centre of the base, and the envelope completed by a glass 'can' sealed to the upturned rim of the base. After processing,
the base and the sealed-off exhaust tube were protected by a perforated aluminium pressing which incorporated a central locating spigot and was held in place
by an aluminium can which covered the remainder of the glass envelope.

The use of 'Loctal' technology extended the upper frequency of operation of
signal amplifiers to several times that of the 'Octal' series, so that high gain low
noise IF amplifiers could be designed to operate at 45 MHz with bandwidths
adequate for the amplification of pulses with durations measured in tenths of a
microsecond. Type Approved as VR91, the EF50 replaced VR65A in British
radar RF and IF amplifiers and in precision time base circuits such as those

developed by F.C. Williams of TRE. [In late 1939, C.W.Oatley and a small group detached from Bawdsey R.S. to work in the Cavendish Laboratory at Cambridge used the Pye signal amplifier module as the IF amplifier in experimental 200 MHz receivers. The 'Pye strip' later became the standard IF amplifier module in most British radar receivers.] RL7, a variant with aligned grids to reduce partition noise (see Appendix 4) and later Type Approved as VR136, surpassed the 'acorn' as a low noise amplifier at frequencies up to 300 MHz. The EF50 was copied and made in large numbers by Sylvania in U.S.A.

7.5 A new generation

The 'Loctal' series was the first to use the single ended 'all glass moulded base ring of pins' fabrication technique. As the war progressed, the increase in the numbers and the complexity of equipments fitted in aircraft, ships and land based vehicles resulted in a demand for them to be made much smaller. This led to the development of a complete range of miniature components, including active devices of all types and particularly general purpose valves, which were used in greater numbers than before.

From 1944, the Loctal and many of the Octal types were gradually replaced by all-glass 'miniature' tubes, initially of American origin and copied by British manufacturers, and later by types developed in UK. The new series was from the beginning accepted as an international standard format. The first tubes had pressed glass bases with seven nickel iron rigid pins about 1 mm diameter and 8 mm long on a pitch circle of about 13 mm. (This required development of sockets which provided reliable contact over time without subjecting the pins to unacceptable mechanical stress). The overall diameter of the cylindrical envelope was about 18 mm. A 9-pin version with an overall diameter of 22 mm was introduced some time later for twin triodes and other multigrid types (for example, hexode mixers), and for power amplifiers where the anode dissipation required a larger area of glass for cooling purposes.

One notable example of this technology was the 6AK5, a high gain signal amplifier pentode designed by Western Electric, which replaced the EF50 as a low noise IF amplifier. Developed in 1945 as a replacement for the EF50, the CV138 also superseded VR65A and other Octal general purpose low power pentodes. It became one of the most widely used British types in this range. [The Octal KT66 survived for many years.]

7.6 Preferred Valves

In the early and mid 1930s, British valve manufacturers produced ranges of 'general purpose' valves in which individual types were marginally different from their competitors. Each maker offered one or more superior characteristics, such as higher gain, lower cathode heating power or more power output from valves of similar size. (Sometimes the difference was only in the static characteristics). By agreement within their trade association, no maker copied a competitor's product. The only common characteristics were the configurations of the valve bases. [In contrast, American manufacturers produced 'vacuum tubes' which were directly interchangeable with those of their domestic competitors. If one manufacturer made an improved version of an established type, all the others were free to make a direct replacement for it immediately it was registered with their trade association and given a type number.] Thus British manufacturers of communications and other specialised electronic equipment tended to select individual types from a number of catalogues in the hope that this would permit them to develop and market units with a performance superior to or a cost lower than that of their competitors. The majority of makers of domestic radio and television receivers bought a 'package' from one or other of the valve makers, in which sets of valves, circuit diagrams and technical assistance were supplied to each customer, often on a confidential basis, in return for an agreement to use only that valve maker's products. There is little or no evidence to suggest that American makers of professional and domestic equipment were handicapped in any way by their use of 'standard' types of tube. So far as the British home market in peacetime was concerned, the diversity of valve types posed no serious problem with the provisioning and holding of spares. In export markets, American equipment manufacturers had the great advantage that a tube with the same type number from any of their suppliers was a direct replacement for others of the same type in many different equipments, which greatly simplified and reduced the range and quantity of spares held by their agents.

In wartime, because of limitations on production capacity and the availablity of raw materials such as nickel, tungsten and mica, all of which had to be imported, and the necessity to minimise holdings of spares in the field, it soon became apparent that the number of different types of general purpose valve in British military equipment had to be reduced. As a result of action by D.N. Truscott, who was responsible for the production of valves for use by Air Ministry and also Secretary of the Interservice Valve Production Committee, Air Ministry and MAP produced a 'preferred list' of about fifty types in September 1940. (see Appendix 2). [Little could be done about equipment in service; the list was intended to limit the number of types used in development of new equipment]. Roughly half of the valves listed were general purpose types. Thereafter the Interservice Technical Valve Committee was given specific responsibility for keeping to the minimum the number of types of general purpose and other valves which they selected for inclusion in a common Preferred List for all three Services, and for its periodic revision, including the preparation of a 'black list'. Although the terms of reference of the CVD specialist committees included responsibility for ensuring the minimum overlap of research and development projects, it was often necessary to investigate two or more possible solutions to a particular operational problem, sometimes as far as the completion of experimental prototypes. A decision then had to be made by TVC (following recommendations from CVD Technical Committee) as to which device was 'Preferred', taking into account performance, ease of manufacture and use, economy in the use of critical materials, and guesstimates of life and reliability. The magnitude of this task, and of the problems associated with providing adequate manufacturing capacity for each type in service, is illustrated by the letter from Truscott to the design

Establishments, which prefaced the third issue of the Preferred List dated June 1941 (see Appendix 2).

VALVE PRODUCTION COMMITTEE
MINISTRY OF AIRCRAFT PRODUCTION,
R.P. Tech. 1
MILLBANK,
S.W.1

List of Preferred Valves
Issue 3

Since the first issue of the Preferred List it has been apparent that a considerable amount of equipment has been designed and made using valves of category II or non-preferred types. Some of this equipment has been adopted for Service use on a small scale by force of circumstance, but as the demands for valves increase it is essential that there be the most strict adherence to the valves in Category I of the Preferred List. If this is not possible for Service Equipment, the Inter-Service Technical Valve Committee must be consulted in order that approval may be given and supplies arranged.

Where equipment is being made for non-Service requirements, e.g. Industrial or Commercial use, it is desirable that the incorporation of any valves not in Category I should be notified to the Inter-Service Valve Production Committee; with an indication of the quantities required, in order that the possibility of supply may be ascertained.

A further list of Transmitting valves is in course of preparation and will be circulated as an addition to this.

D.N. TRUSCOTT,
Secretary

In retrospect, it might be said that so far as general purpose valves were concerned, the arrival of the Octal, Loctal and miniature ranges of receiving tubes (all of overseas origin) was probably the most important factor in the establishment of an effective preferred list, and in causing British receiving tube manufacturers to appreciate that they had to accept international standards if they were to survive in the post war market. The same applied to microwave tubes and modulators, where American expertise had overtaken that available to European manufacturers.

Part 3
Appendices

Appendix 1

**Research
Laboratories of The
General
Electric
Company, Ltd.**

Report to D.S.R.E. Admiralty.
C.V.D.
General Survey of work February to December 1939

2nd January, 1940

The C.V.D. liaison was initiated towards the end of 1938 in order to regularise and extend an informal emergency arrangement, known as U.S.W.A., made in the preceding summer, between D.S.R. and G.E.C. Research Laboratories for the rapid provision of trial valves, other than silica valves, for R.D.F. purposes, the ordinary channels for obtaining such samples having been found to be unsuited to confidential applications.

The initials C.V.D. were intended to denote "communication valves development", not as sometimes interpreted "commercial". According to the original terms of reference commercial considerations were to be eliminated as far as possible, and in fact they have been. For example, contractors, such as Metropolitan-Vickers, have embodied auxiliary valves manufactured by their own associates in lieu of G.E.C. valves used in the earlier stages of development.

The object of the C.V.D. liaison was to assist D.S.R. to keep continuously in review the needs of all three Services and to ensure that valve development was suitably directed or initiated to meet them.

The original demand was for transmitting types, but at an early stage we pressed for and obtained the inclusion of receiving types. At that time, as subsequently, tentative requests for samples of receiving types not yet in production were multiplying.

In later months, as the demands on the transmitting side have approached satisfaction, the receiving side has become more and more prominent. There has been a growing need for correlating experimental trials of samples with the possibilities of early production. Further the view has been gaining strength that there is more to be hoped for from improvement of receiver performance than from enhanced transmitter power.

Finally, there has recently appeared some difference of opinion as to whether the C.V.D. scheme is, or is not, to be considered as confined to the selection or development of valves for R.D.F. purposes. Whatever the original object may have been it would appear that on the receiving side especially, where valves are not so specialized, some consideration of wider fields of application will be inevitable.

Summary of Transmitting Valves, and Uses,
referred to in this report.

The valve type **underlined** is the one at present being used for each application. The others are either alternatives or stages on the way to the ultimate objective.

Signal School	***E. 1046*** (Micropup)	E. 1029 (obsolete)
G.L.	***V.T. 58***	E. 770
Later G.L.	***E. 960T***	
M.B.	***N.T. 77*** (silica)	E. 1024 (Glass-metal tetrode)
		E. 1039 (obsolete)
C.D.	***V.T. 58***	
A.L.)	***D.E.T. 12*** (Low power stop gap).	
A.I.)	***E. 1046*** (now V.T. 90)	E. 981 (obsolete)
		E. 1060 (obsolete)
A.S.V.	***E. 1046*** (now V.T. 90)	

Transmitting Valves.

The general trend of development for special purposes has been in the direction of higher power and shorter wavelength, either or both.

As a first stage, development took the form of minor modifications in design or production technique applied to previously existing types. This has resulted in the adoption of three types which are now in production at rates adequate to existing demands. These three types are:

(a) ***V.T. 58*** (late E. 960) for G.L. and later C.D. down to 1.5 metres wavelength. A longer lived version of the first existing metal-glass valve (E. 770) to be tried out in the summer of 1938 as a shorter wave large production version of the silica N.T. 57. Conversion of the sets for which it was originally designed has latterly considerably widened the scope of this type. Life performance is satisfactory. Production is now on the full scale envisaged and is ahead of contract orders.

(b) ***D.E.T. 12*** (N.T. 58) for A.I. down to 1.5 metres wavelength. A previously existing type somewhat modified in production technique and specially tested at Wembley. Considerable improvement in quality of this type coincided with a revival of demand as a stopgap pending construction of higher power transmitters. The previous difficulties of production on more than a small scale have now been overcome. Complaints of performance in operation are absent.

(c) ***E. 1046*** (micropup) for Signal School down to 50 cm wavelength. For Air

Ministry, A.I. and A.S.V. down to 1.5 metres wavelength. A thoriated filament version of a previous bright filament type E. 1011. This has crystallized out into two versions, one, a very short valve, with a minimum wavelength of less than 50 cm, only suitable for R.D.F. purposes at low altitudes, and the other a longer valve intended for A.I. at altitudes up to 25 000 ft. The latter version, which received special experimental study at Wembley, necessarily has a longer minimum wavelength (ca. 1 metre). For the latter purpose the type is in production on an already considerable scale and first reports on the uniformity and initial performance are satisfactory.

A second stage of development may conveniently be considered as concerned with the embodiment of the thoriated filament in types where bright filaments had previously attained at least a measure of success. The thoriated filament has, of course, long been used in glass and metal-glass valves and the types under (b) and (c) above have it.

The use, however, of the thoriated filament as a new departure in silica valve technique (in the first place in a valve of the N.T. 57 series) accelerated the demand for increased power, life, and/or cathode economy. There are again three types of valve concerned in this stage of the general development. All three are at present undergoing duration tests to establish the stability of the thorium emission.

(d) ***N.T. 77 and E. 1024*** (for M.B. 2. Wavelengths between 5.5 and 12.5 metres). These are equivalent versions in silica and metal-glass respectively. The decision, based on experience with N.T. 57T, to replace the large tetrode N.T. 60 by a thoriated version N.T. 77 made this latter fully equivalent to the companion valve E. 1024 which was being developed as a silica substitute. We have assisted in ensuring that the transmitters now on test are designed to take either. Since each N.T. 77 requires 13 graded glass seals the necessary special glasses have been developed and are in production at Research Laboratories as an independent source of supply. The corresponding seals in E. 1024 are metal-glass on the lines of those used in earlier large pentodes. The grid systems of both types are modified from those of the parent N.T. 60 on the basis of trials made in ***E. 1024*** in order to make full use of the higher emission provided by the thoriated filament.

(e) ***E 960T*** (for higher power G.L.) is a thoriated version of V.T. 58 (E960) requested in the first place as needing less cathode power, and secondly as giving greater output; more than double output has been obtained without re-design, which is apparently a better result than that given by N.T. 57T, which is otherwise analogous.

(f) *Other discarded types.*

In addition to the above six types of accepted transmitting valve which either have been or shortly will be adopted for immediate use, certain other discarded types should, for completeness, be mentioned.

E. 961 a metal-glass triode of similar power to N.T. 60 tetrode.

E. 1029 the first "concentric" type of valve, being a concentric version of D.E.T. 12 but adapted by its construction for use in suitable circuits at about one-half the limiting wavelength; used by Signal School for satisfactory experimental trials, but withdrawn in favour of E. 1046 which gave higher power output and was much simpler to produce.

E. 981 a triode intermediate in size between E. 960 and E. 1046 – used particularly in establishing techniques for the thoriated filament in E. 1046 and E. 960T.

E. 1060 a glass equivalent of E. 981 evading the then expected difficulties in large-scale production of such glass valves (e.g. D.E.T. 12). This was the first "purely R.D.F." valve to be developed. The work was done at a time when it was intended to use 2.5 metres for some services. The valve was satisfactory at this wavelength and even as low as 1.8 metres, but when it was decided later to reduce the working wavelength to 1.5 metres the type was abandoned in favour of E. 1046.

E. 1039 an attempted all-glass version of N.T. 60 silica, discarded as probably unpromising in production, in favour of E. 1024. An all-glass version of N.T. 57 could have been made practicable if pressed but was not demanded as N.T. 57 was at the time considered obsolescent.

A third stage covers the phase of development now begun and involves valves designed throughout for the special R.D.F. operation conditions and not merely modified versions of previous types. These valves are expected to reach shorter wavelengths because they are better adapted to the concentric, or line type of circuit which these wavelengths demand, or alternatively to give 5 or 10 times the power at a given wavelength. They are to have an "inside-out" or "everted" construction with a large cathode system outside and a small anode inside. One specimen, not of the smallest size, has been made. Triodes and tetrodes are both included.

Duration Tests.

As a result of 15 years experience, it is considered at least inadvisable to take anything for granted as far as valve life is concerned. It has therefore been our practice, as a first precaution, to give every sample valve issued to a Service Establishment a run under pseudo-operating conditions for sufficient time to guard against premature failure. This goes a long way to ensure freedom from interruption and delay of experimental programmes.

This, however, is not sufficient, and it has been our aim to arrange with the appropriate establishment for life test under proper working conditions at the earliest possible stage in the history of each individual type. This is particularly important where large scale manufacture is envisaged. In the case of the V.T. 58 life tests, pure and simple, were well ahead of large scale manufacture or use of the valves. More recently facilities were offered and are being used whereby two selected transmitters on actual service run valves supplied for the purpose of pre-determined periods and report to us weekly on performance. It is hoped to arrange similar facilities for other types to augment the routine duration test.

In so far as is possible, extended tests, under laboratory conditions are also carried out at Wembley particularly when any change in design or manufacturing procedure has taken place.

Liaison with Contractors.

We should perhaps mention the collaboration between ourselves and the various contractors whom we meet in connection with the use and proper functioning of the various valves in use.

Receiving Valves.

The general position in receiving valve development in this country a year or more ago differed from the transmitting position in that effort was being concentrated on improving valves for operation at little less than the television wavelength rather than extending the wavelength range downwards as had been done for the transmitters. For experimental purposes "acorn" valves were generally used, frequently American samples, production in this country being on a small scale and retarded by lack of pressure.

At that time there was some tentative study of intermediate types between the television range and the acorns, including copying of foreign types of that sort, and it was beginning to be realised that short wave performance was as much or more dependent on short internal connecting leads in the valve, as on obtaining short electron transit-times. Up to the present time, guidance from the C.V.D. meeting has been to confine its programme to wavelengths down to 50 cms.

Z. 62 (E936) High Conductance Conventional Pentode. In this case retention of the existing methods of construction but with the shortest possible leads did produce a useful valve reaching to one-half the television wavelength. However what has been needed is a new construction with a flat glass or ceramic end to the valve giving much shorter leads.

Z. 62 (E936) On Ceramic Base. Here the electrode system was in no way altered, but the shorter leads obtained by mounting on a ceramic base resulted in a still shorter minimum wavelength.

Other Short Base Valves.

Another solution of this problem in glass has appeared in the form of the E.F. 50 and other Philips Mullard types. A ceramic base of similar general application suitable for moderate scale production has been developed at Wembley and is being applied to a number of types which thereby can use shorter leads. Samples of these are now in demand, including an alternative to the E.F. 50 more quickly producible than a direct copy. Recent tests have shown an early sample of this alternative to be satisfactory.

Deflected Beam Valves.

Another line of development which largely evades the long lead and transit time difficulties uses the deflected beam method of controlling the electron stream.

Radio frequency amplifier valves and frequency-changing or "mixer" valves of this sort are now being tried out. The beam forming system was already well developed and attention under the C.V.D. programme has been focussed on the H.F. electrode design and disposition of leads. If their performance proves adequate these valves should serve to relieve the pressure for large scale production of "acorn" types. "Acorn" production is too difficult to be easily expanded indefinitely, and the fact that the life of acorns, though improving, is none too long for continuous operation only increases the demand still further. Progress in this direction has, it must be confessed, been retarded through the difficulty of obtaining good comparative trials under the experimental conditions at present prevailing in some Service establishments. Efforts are being made to overcome this difficulty by extending the work at Wembley further in the direction of circuit trials than has been the practice with transmitting valves. The equipment, material and personnel, which is needed is however more elaborate in the receiving case.

Valves for Concentric Line Use.

A further line of development, analogous to and associated with the corresponding transmitting case, is being followed using valves directly adaptable to concentric or line circuits, and employing in some cases the everted construction. Early experiments with the standard acorn pentode in concentric circuits showed that this valve was capable of considerably better performance than had previously been considered possible. The first valve of the everted construction was a tetrode "the porcupine", which though promising, is now being redesigned to shorten the wavelength yet further. Other types, diodes and triodes, are now being made. All these types in greater or less measure will raise production problems because of their departure from orthodox receiving valve construction.

Low Noise Level Valves.

Increased attention has recently been devoted to low noise level valves. Samples of a beam tetrode with low screen current but characteristics otherwise similar to Z. 62 (E926), and in which the noise is down by 30%, have been sent to Signal School for test.

Velocity Modulated Valves (Klystron-rhumbatron, etc.)

These valves form a class apart, and are, in fact, in the hands of a separate team at Research Laboratories keeping contact with Signal School and with Birmingham University. Good progress has been made with constructional techniques of a practical sort which have been demonstrated to the meeting.

On the theoretical side, which has had a great deal of attention, careful study of the special relations between the respective properties of valves and of circuit elements resulting from the very high ratios of voltages to currents in the valves, has shown that the most careful prior consideration of experimental conditions is needed if disappointing results are to be avoided. As Oliphant points out this affects, in particular, the relation between power and wavelength in transmitters and the band-breadth requirement in receivers. There seems now to be general agreement between those concerned on these points.

Auxiliary Valves and Devices.

The C.V.D. discussions have naturally brought to light the need for special direction of development for R.D.F. of various auxiliary valves such as high voltage rectifiers, power modulator valves and so forth which need not be separately enumerated although they have been the subject of report and occasionally of intensive effort.

Another form of device which has had attention is the quenched spark gap, in the first place as a means of obtaining an interrupted high voltage supply of suitable wave form and adequate regularity.

This is beginning to look promising and trial units should shortly be built. There has also been discussion with Signal School and others of the use of sparkgaps for the generation of the short wave themselves.

Appendix 2a

                    Issue 1, (amended)
Inter-Service Technical Valve Committee   November, 1941
Service List of Valves for Special Purposes

This Service list of special valves is issued by the Inter-Service Technical Valve Committee for the guidance of designers of equipment intended for certain Special applications. This list is distributed only to Service Establishments and to Contractors engaged in this specific work; it is therefore not available for the designers of communications sets, etc., and valves from it cannot be used for such purposes without the approval of the Technical Valve Committee.

2. Whenever possible, valves shown in the Services List of Preferred Valves should be chosen in preference to the valves in this special list.

3. The majority of the valves in this list are used for certain special applications and many are not suitable for use in circuit arrangements which differ from those for which the valves were specifically developed.

4. Every proposal to use valves in this list for a new application must be submitted to the Inter-Service Technical Valve Committee for approval in order that arrangements for manufacture may be made by the Inter-Service Valve Production Committee.

Issued on behalf of the Inter-Service Technical Valve Committee by the Secretary:

Mr H.G. Hughes
Admiralty Signal Establisment (Extension)
Waterlooville, Hants.

ABBREVIATIONS

O	. . . . .	Obsolescent Type.

Cathode

D.H	. . . . .	Directly Heated
I.H	. . . . .	Indirectly Heated

Base

B5	. . . . .	British Standard 5 pin
B7	. . . . .	British Standard 7 pin
B.O	. . . . .	British Octal.
I.O	. . . . .	International Octal.
T4	. . . . .	Transmitting, 4 pin, metal shell, 47 mm diam.

Cathode Ray Tube

E	. . . . .	Electrostatic
E/M	. . . . .	Electromagnetic

Characteristics

Va	. . . . .	Anode Voltage
Vg1	. . . . .	Control-grid voltage
Vg2	. . . . .	Screen-grid voltage
Ia	. . . . .	Anode Current
G	. . . . .	Mutual Conductance
mu	. . . . .	Amplification factor.
Ra	. . . . .	Anode AC resistance or impedance.

Receiving Valves   Issue 1, November, 1941

Function	Service Type	Cathode	Base	Rating					Characteristics							Commercial Prototype and Remarks
				Cathode		Anode Volts Max	Screen Volts Max	Anode Watts Max	G Ma/V	mu	Ra Ohms	Measured at				
				Volts	Amps							Va	Vg2	Vg1	Ia	
High mu Midget Triode	CV1	D.H.	flex-ible leads	1.35	0.1	–	–	–	0.5	30	60,000	45	–	0	0.3	Mullard DC51 G.E.C. E1177
Triode	NR75	I.H.	B.5.	4.0	1.0	700	–	–	7.0	20	2,900	100	–	0	25	Matched Pairs Mazda AC/P4
Triode	NR94	I.H.	B.5.	4.0	1.0	700	–	–	7.0	20	2,900	100	–	0	25	Mazda AC/P4
Triode	VR117	I.H.	B.5.	4.0	1.0	250	–	–	3.0	45	15,000	100	–	0	3.5	Cossor 41 MT1
Triode	CV6	I.H.	I.0.	6.3	0.2	250	–	3.5	2.0	20	10,000	250	–	−4.7	14	G.E.C. E1148
Triode	CV16	I.H.		4.0	0.65	300	–	–	4.5	100	22,200	300	–	−2	0.5	S.T.C. S25A
Triode	CV24	I.H.	B.O.	4.0	0.65	250	–	–	3.5	36	10,300	100	–	0	8	Mazda HL41
Rugged Midget Pentode	CV4															G.E.C. W1229
Pentode HF.	ARP23 NR70 VR124	I.H.	B.7.	4.0	1.0	200	100	–	2.8	–	–	200	100	−1.5	5	Cossor MS/Pen
Pentode HF.	NR71 VR129	I.H.	B.7.	4.0	1.0	200	100	–	2.8	–	–	200	100	−1.5	5	Cossor MS/PenT
Pentode HF	VR116 (mod)	I.H.	B.O.	6.3	0.6	350	250	–	3.3	–	–	200	200	−4		Mazda V872 Short G3 characteristics
Pentode	CV9	I.H.	B.7.	4.0	2.1	300	275	18	14.5	–	–	250	250	−0.7		Mullard AL60

Receiving Valves   Issue 1, November, 1941

Function	Service Type	Cathode	Base	Rating						Characteristics							Commercial Prototype and Remarks
				Cathode		Anode Volts Max	Screen Volts Max	Anode Watts Max		G Ma/V	mu	Ra Ohms	Measured at				
				Volts	Amps								Va	Vg2	Vg1	Ia	
Variable mu H.F. Pentode	CV21	I.H.	B.O.	4.0	0.65	250	250	–		3.0	–	–	250	200	0		Mazda VP41
Pentode	VT127	I.H.	B.O.	4.0	1.75	330	220	20		8.5			100	100	0	50	Mazda Pen 46
Pentode	NR74	I.H.	B.7.	4.0	1.75	330	220	20		6.5			100	100	0	50	Mazda AC6 Pen

Function	Service Type	Cathode	Base			A1	A2	Grid	Reflector	Input Watts	Output mW	Commercial Prototype and Remarks
V-M Local Oscill-ator Valves	NR89 CV10 CV11	I.H.	I.O.	4.0	1.5	92.5	1850	– 75	– 150	10 max	70 min	Naval Type Army Type Air Force Type
	CV35 CV36 CV37	I.H.	I.O.	4.0	1.4	–	1400	0 to – 50	– 250 to – 320	10 max	100 min	Improved NR89 CV10 CV11

Transmitting Valves   Issue 1, November, 1941

Function	Service Type	Cathode	Base	Rating				Commercial Prototype	
				Cathode		Anode			
				Volts	Amps	Max Volts	Max Watts		
Triode	CV14	D.H.	–	9	60	25 000	100		Silica Valve
Triode	CV15	D.H.	–	3.25	7	4000	1	G.E.C. E1266	
Triode	NT86	D.H.	–	10.5	100	25 000	300		Silica Valve
Triode	NT93	D.H.	–	10 approx	12.0	10 000	100	G.E.C. E1161	
Triode	NT97	D.H.	–	10 approx	12.0	10 000	100	G.E.C. E1161 (modified)	
Triode	NT99	D.H.	–	6.0	8.0	8000	150	G.E.C. E1232	
Triode	VT58	D.H.	–	12.5	60	27 000	700	G.E.C. E960	
Triode	VT90	D.H.	–	8.25	7	10 000	80	G.E.C. E1043	
Triode	VT98	D.H.	–	8.25	35	30 000	700	G.E.C. E960T	Thoriated version of VT58
Magnetron	NT98	I.H.	–	8.0	1.6	9000	200	G.E.C. E1189	Naval Waveband
Magnetron	CV38	I.H.	–	8.0	1.6	9000	200	G.E.C. E1198	Air Force Waveband
Tetrode	CV50	D.H.	–	10.0	70	20 000	300		Silica Valve
Tetrode	VT114	D.H.	–	10.0	70	20 000	200	G.E.C. E1024	

Modulator Valves Issue 1, November, 1941

Function	Service Type	Cathode	Base	Rating					Commercial Prototype	
				Cathode		Anode				
				Volts	Amps	Max Volts	Peak Amps	Max Watts		
Gas Triode	NGT8 (0)	D.H.	—	2.5	42	15 000 (peak)	200		G.E.C. E1191	Wrapped for restriction of heat loss
Gas Triode	CV7	D.H.	—	2.0	11	7000	60	300	G.E.C. E1209	
Gas Triode	CV12	D.H.	—	2.4	44 approx	15 000	200		G.E.C. E1191 High rating	Special fittings for use with auto. temp. control equipt
Gas Triode	CV13	D.H.	—	2.5	20	15 000	100		B.T.H. BT9B	
Gas Triode	CV22	D.H.	—	2.5	20	20 000	50	300	B.T.H. BT45	
Triode	NT78A	D.H.	Flex	10	65	20 000	50	300		Silica Valve
Tetrode	NT100	D.H.	T4	6.0	6.0	12 000		60	G.E.C. E1155	

Cathode Ray Tubes   Issue 1, November, 1941

Service Type	Commercial Type	Maker	Nominal Screen Diameter	Focus Defl.	Final Anode Volts Max	Sensitivity mm/V		Heater		Max. Dimensions (mms)	
						X	Y	Volts	Amps	Diam.	Length including socket
NC1	4053	G.E.C.	$1\frac{1}{2}$"	E.E.	300	120	120	4.0	0.8	40	160
VCR138	4203	G.E.C.	$3\frac{1}{2}$"	E.E.	2000	350	780	4.0	1.0	90	340
NC10	3230K	Cossor	4"	E.E.	4000	600	675	4.0	1.0	135	495
NC12	4201 (mod)	G.E.C.	5"	E.E.	3000	625	1170	4.0	1.0	160	431
VCR112	V1026	Ediswan	$3\frac{1}{2}$"	E.E.	5000	880	300	4.0	0.63	135	495
VCR87	3214	Cossor	6"	E. E/M	4000	700	750	4.0	1.1	165	520
NC7	4602	G.E.C.	12"	E.E.	6000	1625	1270	4.0	1.0	295	638
VCR85	4805	G.E.C.	12"	E.E.	8000	1350	1250	4.0	1.0	305	500
VCR140	–	Cin Tel	12"	M.M.	3000	–	–	4.0	1.1	306	567

Rectifiers   Issue 1, November, 1941

Function	Service Type	Base	Rating				Commercial Prototype	Remarks
			Cathode		Anode			
			Volts	Amps	Volts	Current		
Rectifier half-wave	CV5	Edison 1½" screw	4.0	11.0	20 000 peak inverse	15mA	G.E.C. GU21 Special	Short Ionisation Time
Rectifier half-wave	NU30	Flexibles	14	10.0	10 000 peak inverse 60 000 peak inverse	8 amps peak 800mA	–	Silica, seals at both ends
Rectifier	CV19	Edison screw	16.5	15.0	50 000 peak inverse	1.1 Amps peak	Ediswan V1001	
Rectifier	CV20	B.4	4.0	2.5	4200 peak inverse	600 mA peak 75 mA mean	Ediswan V1900	
Rectifier full wave	CV31	B.4	4.0	3.0	850-0-850	120 mA	U20 and FW4-500 (mod)	

Miscellaneous   Issue 1, November, 1941

Function	Service Type	Base	Rating				Grid Volts	
			Cathode		Anode			
			Volts	Amps	Volts	Current		
Gas Triode	NGT5	B.4	2.5	3.0	1000	2.0 Amps peak 0.5 Amps mean	Trigger Voltage -6 for $V_a = 500$	B.T.H. BT19 Hg filled
Gas Triode	NGT7	American medium 4 pin	5	5.0	1000	12.5 Amps peak 2.5 Amps mean	Trigger Voltage -4 for $V_a = 1000$	B.T.H. BT35 Hg filled
Midget Gas Triode	CV2	Flexibles	1.35	0.15	110	20 mA min	Trigger Voltage -5	G.E.C. E1180 Mullard DAG1
Rugged Midget Gas Triode	CV5							G.E.C. E1228
Midget Discharge Tube	CV17	Flexibles	Striking Voltage 75 to 85 Volts					
Spark Gap	V1.507	Wires	Breakdown Voltage 650 Volts					
Diode Switch	CV8	—	6.3	2.5	240	12	—	G.E.C. E1248 for T and R Aerial

Appendix 2b

Advance Copy
Issue 3, June, 1941
Previous issue should
be destroyed.

Inter-Service Technical Valve Committee.
Services List of Preferred Valves.

This Services list of Preferred Valves has been prepared by the Inter-Service Technical Valve Committee with the object of guiding designers in their choice of valves for use in Service apparatus.

So far as new apparatus is concerned, it will not be permissible in future to use any valve not in Category I of the Preferred List unless prior approval has been obtained from the Inter-Service Technical Valve Committee so that suitable arrangements may be made for production.

The Inter-Service Valve Production Committee will not arrange for the production of a new valve unless the Technical Valve Committee has approved its introduction.

The above restrictions do not apply to the use of valves in small quantities for experimental purposes only, but it should be noted that many commercial types of valves, not in the Preferred List, may be taken out of production at any time.

Further copies of the list may be obtained from:

Mr H.G. Hughes
Secretary, Inter-Service Technical Valve Committee,
H.M. Signal School (Extension),
75, London Road,
Waterlooville, Nr. Portsmouth,

Dr. D.N. Truscott,
Secretary, Inter-Service Valve Production Committee,
Ministry of Aircraft Production,
Millbank, S.W.1

Issue 3, June, 1941
Previous issues should
be destroyed.

Services List of Preferred Valves.

Category I.   Valves which may be used without restriction.
Category II.   Valves, the demand for which should be restricted.
Category III.   Black List. Valves which should not be used.

Note. (a) Although several Service valves may be shown as equivalent, they may differ in specification tolerances.

(b) In some cases, the commercial valves are selected to meet Service specifications; in other cases the Service valve characteristics may differ from those of the prototype.

Reference Table to Abbreviations.

Cathode			Base		Characteristics	
D.H. =	Directly heated	B4	= British Standard 4 pin	G	= Mutual conductance	
		B5	= British Standard 5 pin	μ	= Amplification factor	
I.H. =	Indirectly heated	B7	= British Standard 7 pin	Ra	= Anode A.C. resistance or impedance	
	with separate	B9	= British Standard 9 pin	Va	= Anode voltage.	
	cathode	B.O.	= British Octal	Vg1	= Control-grid voltage.	
		I.O.	= International Octal	Vg2	= Screen-grid voltage	
		UX4	= American 4 pin	Ia	= Anode current	
		UX5	= American 5 pin	E	= Electrostatic	
		US7	= American 7 pin Medium			
		L4	= Special 4 pin Low Loss			
		T4	= Transmitting 4 pin Metal shell			

Changes made to Issue 2.

1. The following valves have been eliminated:

NU17)	G.E.C.	Recommended alternative:	
AU3A)	MU12/14	5Z4G	
VU39)			
NU12)	G.E.C.	Recommended alternative:	
AU1)	U18	5U4G	

It is not intended that sets, recently designed, or in process of being designed, to use either of these valves, should be altered to take the recommended alternatives. Provision of the valves will be made by V.P.C. if the information is given regarding the requirements.

2. The following valves have been introduced:

VCR97	6 ")	
VCR139A	$2\frac{1}{2}$")	
VCR131	12")	Cathode-ray tubes.

3. Type RK48 has been withdrawn pending a final and additional list of preferred transmitting valves now in preparation.

Category I   Issue 3, June, 1941

Function	Service Type See Note(a)	Cathode	Base	Rating				Anode Watts Max	G mA/V	μ	Characteristics				Commercial Prototype See Note(b)
				Cathode		Anode Volts Max					Ra ohms	Measured at			
				Volts	Amps							Va	Vg1	Ia mA	
Diode, double	ARDD5 VR.54	I.H.	I.O.	6.3	0.2	200	–	–	–	–	–	–	–	–	Mullard EB34
Triode	6J5G VR67	I.H.	I.O.	6.3	0.3	250	2.5	2.6	20	7750	250	– 8	9	American 6J5G
Triode	AT20	D.H.	B4	6.0	1.1	600	20	3.2	10	3200	500	– 27.5	40	Mullard MZ05-20
Triode	NR47 VR40	D.H.	B4	4.0	2.0	400	25	7	10	1400	400	– 34	100	Ediswan PP5/400 G.E.C. PX25 Mullard DO24
Triode	AT35 VT25	D.H.	L4	8.0	2.0	1200	36	2.2	10	4550	1200	– 70	30	G.E.C. DET25
Triode	VT30	D.H.	None	12.5	5.5	5000	250	1.35	25	18 600	2000	0	65	
Triode, double	VT61B	I.H.	US7	6.3	0.8	300	7.5	2.8	14	5000	250	– 7	–	American RK34

Category I   Issue 3, June, 1941

Function	Service Type See Note (a)	Cathode	Base	Rating						Characteristics							Commercial Prototypes See Note(b)
				Cathode		Anode Volts Max	Screen Volts Max	Anode Watts Max	G Ma/V	μ	Ra ohms	Measured at					
				Volts	Amps							Va	Vg2	Vg1	Ia mA		
Triode, double diode	6Q7G	I.H.	I.O.	6.3	0.3	250	250	–	–	70	58 000	250	–	– 3	1.1		American 6Q7G G.E.C. DH63
Tetrode beam	6V6G	I.H.	I.O.	6.3	0.45	250	250	12	4.1	–	52 000	250	250	– 12.5	45		American 6V6G
Tetrode, beam	ATS25 VT60A	I.H.	UX5	6.3	0.9	600	300	25	7.1	–	–	300	250	– 12.5	83		R.C.A.807 S.T.C.5B/250A
Pentode H.F.	ARP4	D.H.	B7	2.0	0.1	150	150	–	1.45	–	–	120	120	– 0.5	1.8		Ediswan SP210
Pentode H.F.	ARP36 VR65	I.H.	B.O.	6.3	0.63	250	250	–	8.5	–	–	200	200	– 1.8	8.0		Ediswan SP41 (6.3 V)
Pentode H.F.	6J7G	I.H.	I.O.	6.3	0.3	300	125	0.75	1.2	–	>1.0 Meg	250	100	– 3	2.0		American 6J7G
Pentode variable μ	6K7G	I.H.	I.O.	6.3	0.3	300	125	2.75	1.45	–	–	250	100	– 3	7.0		American 6K7G
Pentode	ATP7	D.H	B7	6.0	0.2	450	250	7.0	3.0 (min)	–	–	450	250	– 5	13		Ediswan V226

Category I   Issue 3, June, 1941

Function	Service Type See Note(a)	Cathode	Base	Rating						Characteristics					Commercial Prototypes See Note(b)
				Cathode		Anode Volts Max	Screen Volts Max	Anode Watts Max	G mA/V	Measured at					
				Volts	Amps					V_a	V_{g2}	V_{g1}	I_a mA		
Pentode	ATP35	I.H.	B7	12.0	0.9	1000	500	35	2.0	1000	200	−33	40	Mullard PV1-35
Pentode	VT104	D.H	B5	6.0	1.3	1250	300	40	3.1	1000	300	−23	40	G.E.C. PT15
Pentode	NT38 ATP75	D.H	T4	10.0	2.0	1500	400	75	1.7	1500	400	−135	50	Mullard PZ1-75
Frequency Changer	6K8G	I.H	I.O	6.3	0.3	300 125	150 (Oscillator Anode)	0.75	0.25 (conv cond.)	1250 100	100 −	−3 Osc.	2.5 3.8	American 6K8G
Frequency Changer	NR82 VR99	I.H	I.O	6.3	0.34	250	100	−	0.2 (conv cond.)	250 100	100 −	−3 Osc.	2.0 4.5	G.E.C. X65
Rectifier Full Wave	5Z4G	I.H	I.O	5.0	2.0	500 (R.M.S. per anode) 125 mA max. rectified current								American 5Z4G
Rectifier Full Wave	5U4G	D.H	I.O	5.0	3.0	550 (R.M.S. per anode) 225 mA max rectified current								American 5U4G
Rectifier Full Wave	VU71A	I.H	I.O	5.0	3.0	550 (R.M.S. per anode) 225 mA max rectified current								G.E.C. U52
Rectifier Half Wave	AU5 VU111	D.H	B4	4.0	1.1	5000 (R.M.S.) 50 mA max. rectified current Peak inverse voltage 12 000								Ediswan V1907 G.E.C. E1132

Category I Issue 3, June, 1941

Function	Service Type See Note(a)	Cathode	Base	Rating and Characteristics					Commercial Prototypes See Note(b)
				Cathode		Anode Volts Max.	Screen Volts	Anode Watts Max.	
				Volts	Amps				
Rectifier	AU6	D.H.	B.4	4.0	2.9	1500 (R.M.S.) 250 mA max.			Mullard RG1-240
Half Wave	VU72					rectified current (gas filled)			G.E.C. GU50
Rectifier	NU33	I.H	B.4	2.0	1.5-	5000 (R.M.S.) 5 mA max.			Cossor SU2150A
Half Wave	VU120				2.0	rectified current			
Rectifier	NU13A	D.H	L.4	6.0	2.0	2000 (R.M.S.) 250 mA max.			G.E.C. U15
Half Wave	AU12					rectified current			Mullard RZ1-250
Rectifier	VU133	I.H	B.4	4.0	1.3	2500 (R.M.S.) 60 mA max.			Ediswan V960
Half Wave						rectified current. Peak inverse voltage 6500			
Voltage	AW3	–	B.4			Operating Voltage 115-135			Cossor S130
Stabiliser	VS110					at 75 mA			
Tuning	NR69	I.H	I.O	6.3	0.3	250 250 Vg for $20°$–$80°$ =			G.E.C. Y63
Indicator	VI103					(Target) 15 volts			

Function	Service Type	Cathode		Screen diam.	Foc. and defl.	A3 Max. Kv	Sensitivity $\frac{mm/V}{X/Y}$		Bulb diam. max. mm	Length max. mm.	
		Volts	Amps								
Cathode Ray Tubes	ACR10 VCR139A	4.0	1.1	$2\frac{3}{4}''$	E.E	1	170	170	70	205	Screen – green
	VCR97	4.0	1.1	$6''$	E.E	3	605	1100	160	431	Screen – green

Category II   Issue 3, June, 1941

Function	Service Type See Note(a)	Cathode	Base	Rating Cathode Volt	Amps	Characteristics				Commercial Prototypes See Note(b)
Diode V.H.F.	VR92	I.H	None	6.3	0.15	V_D (Volts)	− 0.3	− 1	10.0	Mullard EA50
Diode VHF	ARD2 VR78	I.H	None	4.0	0.2	I_D	≮ 5μA	≯ 5μA	≮ 34 mA	Ediswan D.1.
Diode double	VR119	I.H	B.5	4.0	0.75					Cossor DDL4

Function	Service Type See Note(a)	Cathode	Base	Rating Cathode Volts	Amps	Anode Volts Max	Anode Watts Max	G mA/V	μ	Ra ohms	Measured at Ia	Vg1	Ia mA	Commercial Prototype See Note(b)
Triode	VR66	I.H	B.O	6.3	0.64	250	4.0	8.0	17	2100	100	0	35	Ediswan P41 (6.3V)
Triode	VR122	I.H	B.5	4.0	1.0	200	10.0	7.5	11.2	1500	100	0	48	Cossor 41MXP
Triode V.H.F.	NR88	I.H	None	6.3	0.25									Mullard RL18
Triode V.H.F.	VR137	I.H	None	6.3	0.5	300	7.5	7.0	60	8500	100	0	−	Mullard RL16
Triode	VT76	D.H	UX4	7.5	2.5	1000	40	3.6	62	17 500	1000	+ 6	40	G.E.C. DA41 American TZ40
Triode V.H.F.	NT58 NT58A VT62	D.H	B.4	7.5	3.25	1250	50	2.0	10	5000	1000	− 55	50	G.E.C. DET12 RCA 834 STC. 4304BB
Triode	VT26A	D.H	L.4	12.0	1.85	2000	75	1.0	22	22 000	1000	0	100	STC. 4062A

Category II   Issue 3, June, 1941

Function	Service Type See Note(a)	Cathode	Base	Rating						Characteristics							Commercial Prototypes See Note(b)
				Cathode		Anode Volts Max	Screen Volts	Anode Watts Max	G mA/V	μ	Ra ohms	Measured at				
				Volts	Amps							Va	Vg2	Vg1	Ia	
Double Diode Triode	AR8	D.H	B.O	2.0	0.05	150	–	–	1.2	25	21 000	120	–	–1	2.1	Ediswan HL23DD
Triode double diode	NR48 AR21 VR55	I.H	I.O	6.3	0.2	300	–	1.5	2.0	30	15 000	250	–	–5	6	Mullard EBC33
Tetrode beam	6L6G	I.H	I.O	6.3	0.9	350	250	19	6.0	–	22 500	250	250	–14	72	American 6L6G
Tetrode	ATS250 VT31	D.H	None	11.25	8.0	5000	1000	250	1.0	100	100 000	3000	600	–35	82	Mullard SG250
Pentode H.F.	VR56	I.H	I.O	6.3	0.2	300	125	1	2.0	5000	2.5 Meg	250	100	–2	3.0	Mullard EF36
Pentode H.F.	ARP35 VR91	I.H.	Spec-ial	6.3	0.3	300	300	3.0	6.5	–	–	250	250	–2	10	Mullard EF50
Pentode Variable μ	ARP12	D.H	B.O	2.0	0.05	150	150	–	1.15	–	–	120	60	–0.5	2.3	Ediswan VP23
Pentode Variable μ	ARP34 VR53	I.H	I.O	6.3	0.2	300	300	2	2.2	–	–	250	100	–2.5	6.0	Mullard EF39
Pentode V.H.F.	VR136	I.H	Spec-ial	6.3	0.3	250	250	–	7.7	–	–	250	250	–2	10	Mullard RL7
Pentode Output	NR43	D.H	B.5	4.0	0.25	300	200	6.0	1.6	–	–	150	150	–2.5	33	Mullard PM24A
Pentode Output	VT52	I.H	I.O	6.3	0.2	300	250	7.5	2.85	–	–	250	250	–18	30	Mullard EL32

Category II   Issue 3, June, 1941

Function	Service See Note(a)	Cathode	Base	Rating					Characteristics						Commercial Prototypes See Note(b)
				Cathode		Anode Volts Max	Screen Volts Max	Anode Watts Max	G mA/V	μ	Measured at				
				Volts	Amps						Va	Vg2	Vg1	Ia mA	
Pentode	ATP4	D.H	B.O	2.0	0.3	150	150	4.0	3.6 (min)	–	150	150	–8	38	Ediswan V248A
Pentode	NT65	D.H	T4	4.0	2.0	1000	300	35	1.5	–	1000	250	–65	35	Mullard PZ1-35
Pentode	ATP100	D.H	American Giant 5 pin	10.0	5.4	2000	400	100	5.25	–	2000	400	–20	50	S.T.C. 4069A
Frequency Changer	ARTP1	D.H	B.9	2.0	0.26	150	150	(Pen:)	1.3	–	120	60	0	2.2	Ediswan TP22
						150	–	(Tri:)	1.3	33	100	–	0	2.0	
Frequency Changer	ATRP2	D.H	B.O	2.0	0.2	150	150	(pen:)	1.0	–	120	60	0	2.6	Ediswan TP25
						150	–	(Tri:)	1.7	18	100	–	0	5.3	
Frequency Changer	ARTH2	I.H	I.O	6.3	0.3	300	150	(Hex:)	0.65 (Conv.		250	100	2.0	2.0	Mullard ECH35
						150	–	(Tri:)	Cond.)		100	–	Osc.	3.3	
Rectifier Full Wave	6X5G	I.H	I.O	6.3	0.6	350 volts (R.M.S. per anode) 75 mA Max. rectified current.									American 6X5G
Rectifier Half Wave	AU7	D.H	Goliath Edison Screw	4.0	15.0	2000 Volts 750 mA mean rectified current Peak inverse voltage 10 000 (gas-filled)									Ediswan ESU.300 Mullard RG3/1250 S.T.C. 4049C

Category II   Issue 3, June, 1941

Function	Service Type See Note(a)	Cathode	Base	Rating		Anode Volts Max.	Characteristics	Commercial Prototypes See Note(b)
				Cathode				
				Volts	Amps			
Gas Triode	NGT2 VGT128	I.H	B.5	4.0	1.3	500	Control Ratio 28 Voltage Drop 16V	G.E.C. GT1C
Voltage Stabiliser	AW4 VS68	–	B.5	–	–		Stabilised voltages: 285, 214, 142, 71 volts	G.E.C. STV. 280/40

Function	Service Type	Cathode		Screen diam.	Foc. and defl.	A3 Volts KV. max	Sensitivity Mm/V / X/Y	Bulb diam. max. mm	Length max. mm.	Commercial Prototypes See Note(b)
		Volts	Amps							
Cathode Ray Tube	VCR131	4.0	1.1	12"	E	4	600 600	295	585	Cossor 3241 Green screen

"Acorns":	Admiralty	Type NR50, NR54, and NR54A.
	Air Ministry	Type VR95 and VR95A.
	Air Ministry	Type VR59
	G.E.C.	Type HA1
	G.E.C.	Type HA2
	G.E.C.	Type ZA1
	G.E.C.	Type ZA2
	Mullard	Type AT4
	Mullard	Type AP4
	Mullard	Type 4671
	Mullard	Type 4672
	Ediswan	Type A40

"Giant Acorns":	Standard Telephones &	Type 3B/250A
	Cables	Type 4316A
"Micromesh" Valves:	S.T.C.	Type 4033A
	Admiralty	Type NT37

Appendix 3

S.R.E. Department,
Admiralty.
January 1942.

C.V.D.

The following notes present in bare outline the activities of the Coordination of Valve Development organisation from 1939 to the present date.

1939

The first C.V.D. Policy meeting was held in November of this year and it was then that the C.V.D. organisation in its present form began to take shape. Subsidiary to the Policy Committee were the Technical Committee with Service representation only and the G.E.C. liaison committee in which G.E.C. representatives took a large part.

1940

The subject of Coordination was very fully discussed at a meeting of the C.V.D. Policy Committee held at the Admiralty in March of this year.

The Committee decided that coordination of research and development work on valves for all purposes should be dealt with on the same lines as for R.D.F. transmitting valves and recommended that the responsibility for this coordination should rest with D.S.R. Admiralty acting through the C.V.D. Policy Committee. This proposal was subsequently agreed to by the Services. Professor R. Whiddington F.R.S. came to the Admiralty in May 1940 and spends his whole time on C.V.D. matters.

During this year considerable expansion of the C.V.D. machinery took place since it was found that the work on valves was proceeding so rapidly that it was necessary to set up special committees to deal with the various sections.

The new committees set up were:

The 50 cm. Committee
The 10 cm. Committee
The Modulation Technique Committee.

Each Committee has a chairman drawn from the Ministry most interested in the subject matter of the Committee, and a Secretary chosen by him. These committees report to the Policy Committee. From time to time as necessity arises special ad hoc committees are called.

1941

Further expansion of work took place in this year which led to the widening of the ambits of the 50 cm. and 10 cm. Committees. Each Committee found that the range of wave lengths dealt with was widening and to indicate this their names were changed from 50 cm. to Decimetre; and from 10 cm. to Centimetre. The former Committee now deals with all valves ranging from 150 cm. down to 25 cm., while the Centimetre Committee deals with all wave lengths below this. All this time the communications valves had been dealt with partly by C.V.D. Technical Committee, and partly by the Communications Committee of the W/T Board. The work of this latter Committee was transferred to the C.V.D. organisation in this year.

During this year also further development contracts were placed for C.V.D. with firms other than the G.E.C. The S.T.C. contract, for example, was taken over; a contract with the B.T.H. for special modulator valves, and later for valve developments generally, was placed. Contracts were also placed with Messrs. E.M.I., and Messrs. Cosmos.

It had been realised, in setting up a Committee to deal with modulation valves, that the Committee would have to concern itself necessarily not only with valves, but other methods of modulation and with certain ancilliary apparatus. For this reason the title "Modulation Technique" was applied to this Committee and certain special contracts have been placed, notably with Messrs. Metropolitan-Vickers, for development of pulse transformers. Quite apart from the six firms already mentioned, and the Service Establishments research and development are going on at the following centres:

University of Birmingham (Physics Department)

Professor Oliphant and his team are developing magnetrons and other valves in the centimetre region.

University of Birmingham (Electrical Engineering Department)

Professor Dannatt is investigating the magnetic properties of ferro-magnetic cores for pulse transformers.

University of Sheffield

Professor Sucksmith is investigating the stability of permanent magnets used for magnetrons.

University of Manchester

Professor Hartree is investigating the theory of the magnetron.

University of Leeds

Professor Stoner is also investigating the theory of the magnetron.

University of Oxford (New Clarendon Laboratory)

A team of workers is investigating new type valves and problems arising out of them, mainly in the centimetre region.

University of Bristol

Professor Tyndall's team is giving assistance from time to time to the A.S.E. Group housed in his laboratory.

The C.V.D. organisation is fully mindful of the necessity for cooperation between firms, Universities and Establishments, an object which is attained by encouraging personal visits both ways as well as by the attendance of selected individuals at the C.V.D. Committees relevant to their work.

Coordination emerges to a great extent from the work of these Committees and questions of priority of work may be referred to the Policy Committee which is in a position to take suitable action through its Chairman (D.S.R. Admiralty) in cases of difficulty.

Headquarters is not only the focus of these various activities but is also able to assist the work in such ways as disseminating reports from this country and abroad, maintaining contracts between this country and U.S.A. and the Dominions, assigning priorities in the delivery of experimental valves and arranging contracts for experimental development and production.

In regard to programme no rigid differentiation as between the work – say – of the Universities and the firms is insisted upon, although it is true that some of the pioneer work and much of the long term work has naturally been undertaken by the Universities. Coordinated cooperation is at all times attempted and is frequently a spontaneous expression.

At the present moment, for example, the cavity magnetron (Birmingham originated by Boot and Randall) is being actively pursued into the lower centimetre reaches by joint action between Birmingham and Messrs. B.T.H. at Rugby. The strapping technique (discovered by Sayers at Birmingham only a few months ago) is being applied to magnetrons at both B.T.H. and G.E.C. – the three centres being in the closest collaboration and holding frequent ad hoc meetings to discuss minutiae. Other examples could be cited.

The Broad division of work is as follows:

G.E.C.

Transmitter triodes.
Hard and soft Modulators.
Receiver valves and diodes at short waves.
Magnetrons – split anode and cavity – particularly at 10 cm.
Spark gaps of numerous types.
Crystals – particularly the new non-heating type. [Note: this is probably a reference to high burn-out crystals.]

B.T.H.

High power and 3 cm. magnetrons.
Transmitting and Local Oscillator Klystrons. (C.S. construction)
Thyratrons.
Crystals.

S.T.C.

Tunable valves in the cm. region.
Power oscillators at 3 cm.
Grounded grid valves (10, 50, 150 cm regions).

E.M.I.

Velocity modulated local oscillators.
Velocity modulated mixers.

Cosmos

Hard modulator valves.
Diode rectifiers.
Receiver valves.

Metropolitan Vickers

Pulse transformers for use with magnetrons.

Clarendon Laboratory, Oxford

Klystron type oscillators at 1 and 3 cm.
Magnetrons at 1 cm.
Crystal investigations.
(Atmospheric absorption of very short waves).

Birmingham University, Physics Department

Klystrons.
Megawatt magnetrons and wave guides (10 cm.)
Strapped magnetrons generally.
Short wave magnetrons (3 cm.)
Modulation.

Birmingham University, Electrical Engineering Department

Properties of ferro-magnetics under short pulse conditions.

Manchester University and
Leeds University

Theoretical investigations into the action of the magnetron.

Admiralty Signal Establishment Extension, Bristol University

Power Klystrons.
3 cm. V.M. Local Oscillators.
A.F.C. Local Oscillators.
Skiatrons.

Admiralty Signal Establishment Extension, Liss

This Extension was set up in order to take care of the life testing of valves for the three Services.

Two main objects are in view:

(1) To life test valves in an early stage of development.
(2) To life test valves in production and pre-production.

The valves are mainly tested in sets – in prototype or in finalised form – provided by the interested Service.

The Establishment is already too small for its work and is about to be extended.
There are serious difficulties in getting additional staff.

Admiralty Signal Establishment Extension Waterlooville

A small group dealing with
Silica valves for medium waves.
Rectifier valves.

another group, versed in various valve techniques is always ready to make up special valves for urgent service requirements.

Admiralty Signal Establishment Extension, Great Baddow

This extension deals in the main with circuit problems put up from Haslemere but there is one officer engaged on valve research (communications valves in the lower reaches) and another looking after pre-production of 10 cm. magnetrons.

All the above work is concerned with research and development only, but as it is all aimed at ultimate production, contacts with the production side (M.A.P.) have recently been formalized and are increasing.

There is for example the pre-production committee of two (one from C.V.D. another from D.R.P.) charged with the duty of arranging "pre-production" of such valves as are in early small demand, after having emerged from the development stages. Other examples of such contacts between C.V.D. and the production side is when a C.V.D. valve not yet at the production stage but still in the designers' hands is known to be called for in large numbers.

In such a case a meeting between members of C.V.D., D.R.P. and suitable firms is called so that matters of differences in manufacturing technique can be discussed and adjusted in good time.

<h1 align="center">Appendix 4</h1>

Noise and Noise Factor

Noise: A conductor (or a resistor) comprises equal numbers of positive ions and electrons per unit volume. When a voltage is applied between the ends of the circuit, the current flowing through it is carried by electrons moving from one end to the other, changing their direction at random as they approach the positive ions, which vibrate about their equilibrium positions with an excursion proportional to the absolute temperature of the circuit. Thus at any given instant there will be a difference between the number of electrons entering and leaving it. This varies at random with time, so that a 'thermal noise' voltage is added to the input. Noise has a very wide frequency spectrum, and so the amount of noise power added to the power flowing through the circuit is proportional to its frequency response or 'band width'.

Because the cathode in a thermionic valve operates at high temperature ['red heat', about 850°C, equivalent to 1100°K, the absolute temperature], electrons leave it with a considerable random variation in numbers and velocity, so adding 'shot noise'* to the steady current reaching the anode when a positive voltage is applied to it. Being a thermal effect, the shot noise is in theory directly proportional to the absolute temperature of the cathode. This is true only if the electron emission is 'temperature limited'. In practice the negatively charged electrons which have recently left the cathode influence the departure of those which follow, so that the cathode is surrounded by an electron cloud or 'space charge' in which the electrons lose some of their random characteristics. This 'space charge smoothing' reduces the shot noise added to the anode current.

*[The random arrival of electrons at the anode is analogous to raindrops or solid particles falling on a roof or other solid surface, so that 'shot noise' is an apt description.]

In triodes and tetrodes, the use of a negatively biassed grid to vary the number of electrons reaching the anode increases the density of the electron cloud surrounding the cathode so that the emission is 'space charge limited', which enhances the smoothing effect. In tetrodes and pentodes, this reduction in shot noise is offset by 'partition noise', which is generated by the fraction of anode current interrupted by the second grid. This produces a noise voltage which acts on the anode current, so causing an increase in the noise current carried by the beam. This effect can be minimised by accurate alignment of the grids, so that almost all the electrons passing through the first grid reach the anode without being intercepted by the second. The amount of noise in all thermionic valves is proportional to the anode current. Like thermal noise in passive circuits, the noise power added to the output signal is proportional to the bandwidth of the anode circuit.

Noise Factor (noise figure)

Signals entering an amplifier system will be contaminated by thermal noise originating in the input circuit and shot noise generated in the first valve of the amplifier chain. (The following stages add noise, but to a much larger signal, so that their effect is negligible if the first stage has sufficient gain). The 'noise factor'* of a high sensitivity receiver limits its ability to detect very small signals because it determines by how much the noise content of an incoming signal is increased by electrical noise originating in the first stage of the receiver itself. As its amplitude is reduced, the input signal will ultimately be swamped by the 'front end' noise, which produces the background 'hiss' in radio receivers and the 'snow' on the screens of television sets. This is reduced effectively to zero by automatic gain control when the receiver is tuned to a sufficiently large incoming signal. The noise factor N is independent of the gain of the amplifier and has to be measured indirectly. It is defined by the equation:

$$N = \frac{\text{Output Noise Power}}{\text{Output Signal Power}} \Bigg/ \frac{\text{Input Noise Power}}{\text{Input Signal Power}}$$

It is usually expressed in decibels ($10\log_{10}N$). If $N = 1$, the receiver has introduced no extra noise.

Because noise power is proportional to bandwidth, a highly selective receiver will have a better noise factor than a broad band system. This advantage will be offset by its inability to amplify the high frequency components of a short pulse waveform. Thus the designer of radar receivers has to achieve an acceptable compromise between these two phenomena. Because shot noise is proportional to anode current, the first amplifying stage of the receiver should give high gain with the minimum possible anode current. If it is a mixer it should operate with no more current than that compatible with minimum conversion loss. (see 4.4)

The 45 mHz IF amplifiers used in wartime microwave radar equipment had noise figures in the range 2 to 4 dB. The crystal mixer had about 6 db frequency conversion loss. It also introduced thermal noise. These effects, and the injection of shot noise by the local oscillator caused degradation of the overall noise figure of the receiver, which was typically in the range 7 to 14 dB, so making the signal to noise ratio unity with inputs of the order 10^{-13} W.

*The first reference to the use of 'Noise Factor' as the criterion of receiver performance appears to be a memorandum by J.E. CLegg of TRE and E.G. James of GEC to the CVD '50 cm' subcommittee, dated 28 February 1941. The first published paper referring to Noise Factor was "The Absolute Sensitivity of Radio Receivers" D.O. North, RCA Review, 1942, 6, 0 332.

Noise in Oscillators

An RF self oscillator comprises a power amplifier with a resonant output circuit from which a small fraction of the output power is fed back to the input circuit and provides the drive required to produce that output power.

Shot noise in the anode or the beam current of the amplifier provides the means for initiating the oscillation because at some point in its frequency spectrum there will be a fluctuation corresponding with the resonant frequency of the output circuit. This generates an output voltage of which a small fraction is returned to the input circuit, amplified, and added to the original output voltage. Provided there is sufficient gain in the amplifier and there is positive feedback (i.e. the returning voltage is in phase with that across the input circuit), the output voltage will increase until its amplitude is limited by the operating conditions of the amplifier. In triode oscillators, the output power is limited by arranging that the input power to the grid produces a bias voltage which reduces the amplifier gain as the input voltage increases. In velocity modulation systems, such as the reflex klystron, the output power is limited by the mutual replusion of the electrons in the beam, which controls the amount of bunching that can be impressed upon it.

Because the input to the mixer from the local oscillator is many times greater than the incoming signal, its noise content is a major factor in determining the overall noise factor of superheterodyne receivers when there is no pre-amplification of the input signal to the mixer. Thus local oscillators must operate with the minimum current (to limit shot noise) and a high-Q (high selective) output circuit to minimise the noise content of their output power (see 4.3).

Appendix 5

Line Modulators

The Basic Design

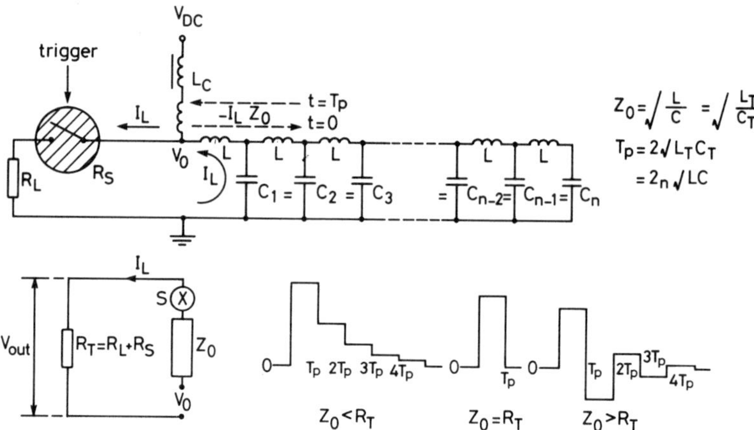

Fig. 1 Line modulator

If a coaxial transmission line is charged to a voltage V_O and then connected to a load R_T, it behaves as a generator with an internal or 'characteristic impedance' Z_O. Maximum power is transferred to the load when $R_T = Z_O$. Under these conditions, a constant current $I_L = V_O/2R_T$ flows into the load. At the same time, a step voltage waveform $-V_O/2 = I_L Z_O$ travels to the open end of the line, where it is totally reflected and returns to the loaded end, so reducing the line voltage to zero and terminating the current into the load. Transmission line theory shows that the time taken for the double transit (the pulse duration) is $2\sqrt{L_T C_T}$. The characteristic impedance $Z_O = \sqrt{L/C}$, where L and C can be defined either as the inductance and capacitance per unit length of the line or as the total inductance L_T and capacitance C_T. [If the line has very low attenuation, it can be assumed that Z_O has the dimensions of a pure resistance].

This is the basic principle of operation of line modulators, invented by O.L. Ratsey of Signal School in 1940. In practice the transmission line is replaced by an iterative network with the total inductance and capacitance required to generate a pulse of specified duration into a load of known impedance.

The Line Modulator

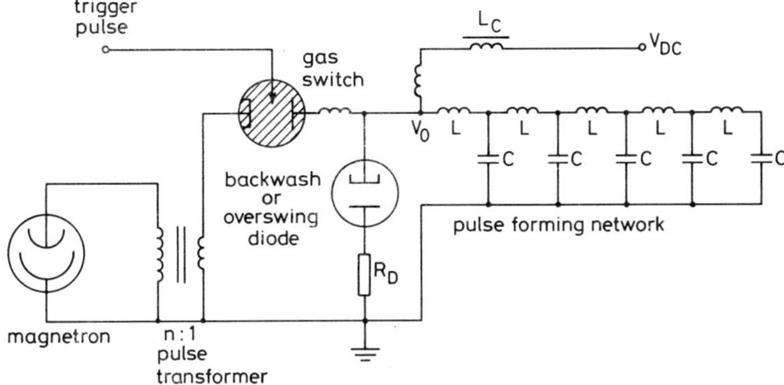

Fig. 2   Line modulator with pulse
transformer and overswing diode

Fig 2 shows a simplified arrangement of a typical line modulator. Energy is stored during the pulse interval by charging the total line capacitance C_T through the inductance L_C, which by resonating with C_T makes the line voltage V_O twice the DC input when the natural frequency of the charging circuit is half the pulse recurrence frequency. [Note: When the pulse recurrence frequency is the same as that of the AC power supply (typically 50 or 400 cycles per second), the network can be charged through a step-up transformer during each positive half cycle, and discharged at the peak]. To minimise distortion of the output pulse by the impedance of the transformer secondary winding or the charging choke, the line is isolated from it by an RF choke (a '10's microhenry inductor with low self capacitance)].

When the trigger pulse is applied to the switch, it initiates a self-maintaining discharge through which power is fed to the load R_L by a constant current $I_L = V_O/(Z_O + R_T)$ for a time $T_p = 2\sqrt{(L_T C_T)} = 2n\sqrt{LC}$. When R_T equals Z_O the line voltage falls to zero at the end of the pulse, the switch discharge is extinguished and the charging cycle is repeated. [Virtually no power is dissipated in Z_O because the line is an assembly of almost purely reactive components]. If there is a mismatch between them, the voltage falls to a fraction of V_O, going negative if R_T is less than Z_O. The voltage across the line then returns to zero in a series of steps as shown in Fig 1. In either case power is lost from the main pulse.

Because of their high conductance and low voltage drop, gas-filled devices such as thyratrons and trigatrons can be used efficiently with low impedance loads and lines. This reduces the supply voltage and so the volume and weight of the modulator unit. [Some RAF equipments used rotary spark gaps in air. In the Naval Type 274 and 275 500 kW S-band radars, triggered spark gaps in air were used in place of mercury thyratrons to obtain rapid start up and permit operation with half microsecond pulses. They were also used to obtain rapid start up in high power land based systems]. In most cases a pulse transformer is used to raise the output voltage to the operating level of the magnetron or other power generator to which the switch is connected. This and the use of a pulse transformer limits the rate of rise of the output voltage and causes some overswing and oscillation of the line voltage after the main pulse is complete. If an arc occurs in the transmitter, this effectively short-circuits the load and produces a large overswing pulse. This may cause reverse conduction in the switch and the load. The network voltage then oscillates about zero until the arc is extinguished. This results in shortened life of the tubes or complete breakdown of the equipment, because the stored energy must be dissipated in the valve and components of the modulator and not in the load. High power line modulators therefore incorporate an overload protection device which disconnects the DC supply when the load current reverses or rises to the danger level. [The energy supplied through the charging inductance may be sufficient to maintain the arc indefinitely.] If the arc is extinguished when the line voltage overswings because the switch or load will not conduct in the reverse direction, the negative voltage left on the line increases the current through the charging inductance, causing the positive voltage to be above normal when the switch is triggered by the next control pulse. This increases the probability of another arc and continuation of the process and early failure of the equipment. A 'backwash diode' is therefore connected across the line so that energy is absorbed by the matched load R_D when the voltage reverses, so causing it to return to zero at the end of the overswing pulse. In practise it is difficult to make the forward resistance of the diode alone equal to Z_L, and so R_D is omitted. The voltage then returns to zero in a few steps of duration T_p.

Parallel Charge — Series Discharge System

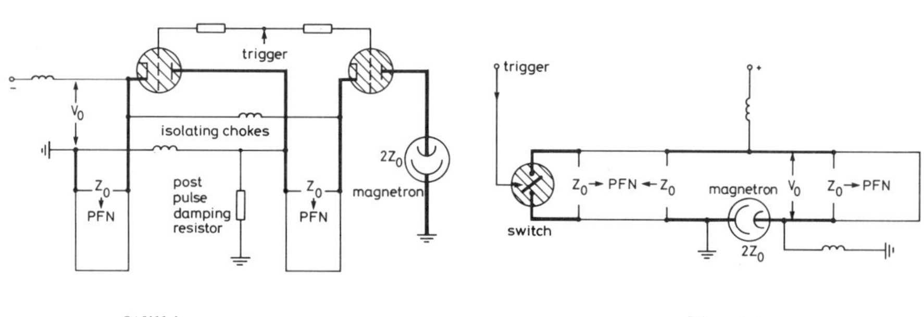

Wilkinson   Blumlein

In 1940, when the first line modulators were developed, no technology had been established for the design and construction of high power pulse transformers. Because the operating voltages of high power magnetrons (from 15 to 25 kV) were too high for the hold-off capability of available mercury thyratrons [this had to be approximately twice the magnetron voltage], systems were developed in which two lines were charged in parallel and discharged in series into the load.

   The first of these, devised in late 1940 by K.J.R. Wilkinson of BTH used two thyratrons to discharge the twin pulse forming networks. A major disadvantage of this system is that one of the switching tubes will be subjected to a large overvoltage if it fails to turn on within a fraction of a microsecond after the other is conducting. Because it is difficult to ensure simultaneous 'firing' of the two switches, particularly over time, this can cause unacceptably short life.

   An alternative system, invented by A.D. Blumlein of EMI in 1941, used a single switch to discharge the two networks. Its basic principle of operation is the same as that of single line systems. When the switch is closed, the voltage at that end of the line is reduced to zero and a step current waveform travels away from it so that the line is charged to $-V_O$ by the time $t = \sqrt{(L_T C_T)}$ at which the wave reaches the open end. The system then functions as two line modulators in series with an output voltage V_O generating a constant current into the load for a time $T_p = 2\sqrt{L_T C_T}$. For maximum transfer of power the total impedance of the load must be $2Z_O$. The delay of half a pulse length between the closing of the switch and the beginning of the pulse simplifies the design of radar monitoring and ranging displays. The Blumlein circuit eliminates the problem of synchronising the initiation of the discharge in thyratrons or trigatron switches in the other twin line system. It has the disadvantage that the switch has to have twice their current carrying capability.

Appendix 6

Materials and technology

(Extracts from 'Electron Physics and Technology', J. Thomson and E.B. Callick, English Universities Press, 1959)

Note: the technologies described are those in established use before 1959, and were used to make the devices described in Chapters 4 to 7 of this book. The subsequent development of semiconductor devices has made some of the techniques obsolete.

1 Cathodes and Emitters

Introductory

An electron device will operate correctly only for the time that the cathode emits electrons in the quantity required by the interaction process. The cathode must therefore have long life at the temperature giving the necessary emission, and the emission be obtained without adverse effect on the properties of other parts of the electrode system, or on some special characteristic of the complete device. Copious thermionic emission is of little use if evaporation of active material modifies the control characteristics of a grid or causes electrical leakage across insulators. It is therefore essential to allow the greatest possible margin between operating current and the saturated emission available under limit conditions, i.e. where the temperature is raised to the point at which evaporation of the coating becomes excessive. At the same time, physical limitations on the size of the device set a lower limit to the operating current density. For example, high transconductance tubes used as amplifiers in wide-band systems require high current density if the advantage of the high slope is not to be offset by the capacitance between grid and cathode. Microwave power valves, being necessarily limited in size, can give their maximum output only by operating with the highest permissible cathode loading. In these circumstances, long life can be obtained only by producing high activity in the cathode coating and minimising the possibility of poisoning during operation. Thermionic cathodes are the source of emission in nearly all the valves made at the present time. The only notable exceptions are magnetrons where secondary emission provides the greater part of the operating current.

Thermionic cathodes

Emission of electrons from a solid surface varies with temperature according to the equation

$$I_s = AT^2 \exp\left(-\phi/kT\right) \tag{1.1}$$

where
   I_s is the saturated emission per unit area in A/cm^2
   T is the absolute temperature in $^\circ$K
   ϕ is the work function
   A and k are constants
   ϕ is a measure of the work done by an electron escaping from an emitter surface, and is expressed in 'electron volts'
   Assuming A to be typically 40 A/cm^2, the saturated emission from surfaces of

different work function at temperatures up to 2500° K is as shown in Fig. 1.1. High current density is obtained only when ϕ is low, and as indicated by the dominance of the exponential term in equation (1.1), the variation of I_s with T and ϕ/T is extremely rapid. Because they require less heating power, the most efficient cathodes are those operating at low temperature. A low work function is therefore essential if the emitter is to be used economically in electron devices operating at low power levels, e.g. in radio receivers and communications equipment.

It has been known for many years that a surface layer of foreign atoms or molecules can modify the work function of a metal by a large amount. For example, barium on tungsten gives a work function of 1.6 V, and thorium 2.7 V.

Fig. 1.1 Theoretical variation of thermionic emission with work function and temperature

Temperature (°K)		1,000	2,000	2,500
Material	ϕ (eV)	Saturated emission per cm^2		
(BaO, CaO, SrO)Ba (oxide cathode)	1.5	1 A		
Ba on W	1.6			
	2.0	3 mA	1400 A	
Ba	2.5	12 μA	80 A	
Th on W	2.7			
	3.0	0.03 μA	43 A	200 A
	4.0		13 mA	2.5 A
W	4.5		0.9 mA	270 mA

At 1900 °K a thorium on tungsten cathode will give many times more emission than a pure tungsten cathode at 2500°K, have a similar life in terms of evaporation of active material and require about one third of the heating power. The oxides of barium strontium and calcium when mixed together have a work function of 1.5 eV, so making oxide coated cathodes the most efficient of all.

Thoriated-tungsten cathodes

The modern thoriated-tungsten cathode contains up to 2% of thoria, the amount being limited by workability of the compound material. Like tungsten, it is freely available only as rod or wire, and so is used in a similar way. Thoriated-tungsten cathodes operate by maintaining a balance between loss of thorium from the surface and diffusion of thorium outwards from the body of the material. If a pure tungsten-thoria combination is used, thorium can be formed only by reduction during the activation at about 2700° K. At operating temperatures around 2000° K, the adsorbed film can be maintained only if the rate of loss is low. This requires very low residual gas pressure, for the thorium can be removed by chemical action and sputtering caused by ion bombardment, as well as by evaporation. To minimise this effect, the filament is *carburised* by maintaining it at a temperature around 2200° K in a stream of hydrogen which has bubbled through benzene or xylene. This produces a surface layer of tungsten carbide, the process being controlled by measuring the fractional increase in resistance of the filament, which is an almost linear function of the thickness of the carbide layer. To maintain adequate strength, the carburising is halted when the thickness of the skin is about one-tenth of the radius. Tungsten carbide reduces thoria at temperatures in the operating range of 1900–2000° K, and so the surface film can be maintained for a very long time under vacuum conditions which would not be tolerated by an uncarburised filament. Even so, the emission from a thoriated-tungsten cathode is more sensitive to gas pressure than that from an oxide emitter. An incidental advantage of carburising, which increases thermal emissivity by about 20%, is that the rate of evaporation of thorium is about one-sixth of that from pure tungsten. The process may be performed either before or after assembly of the cathode into the valve. At 2000° K with 30 W/cm^2 radiation loss, the saturated emission is 3 A/cm^2, or ten times that of tungsten at 2500° K with 70 W/cm^2. Thoriated filaments can therefore be used economically in much smaller devices. Taking the saturated emission requirement to be 2 A, this can be obtained from a filament 10 cm long and 0.2 mm diameter with 20 watts heating power at about 6 volts. At 1900° K, a slightly larger filament giving the same emission would require 25 watts, but have a much longer life because of decreased evaporation. In both cases the heating power is about one-twentieth that of a larger tungsten cathode giving the same total emission. In practice, lives of tens of thousands of hours are obtained with operating current densities up to 1 A/cm^2. Thoriated-tungsten cathodes are used in transmitting valves operating at frequencies up to a few hundred MHz with applied voltages up to 15 kV. In the larger types, they comprise a cage of vertical rods suspended from the seals. The heating current flows in opposite directions through adjacent members, and stiffening spiders are attached at the centre of each set to minimise distortion during operation caused by warping of the rods at high temperature. In the smaller tubes they are usually a directly heated single-helix or a multiple-helix arrangement supported on a centre post.

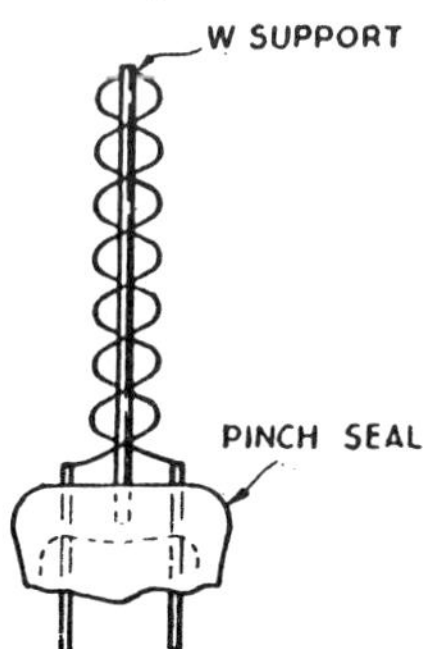

Fig. 1.2   Thoriated tungsten filament

Oxide-coated cathodes

Oxide-coated cathodes are the most widely used of all, being fitted universally in general-purpose valves for communication and radio receiving equipment, and telephone apparatus. (This was the situation until the 1960s. Modern equipment uses semiconductor devices.) The coating is either a mixture of barium and strontium oxide or of these with a smaller amount of calcium oxide on a metal surface to which they are attached by sintering at the activation temperature. Because the heat of formation of barium oxide is lower, the outer surface of a double mixture is a layer of strontium oxide containing an excess of barium either within it or on the surface, and it is this excess of barium which determines the emissive properties. The oxide therefore operates in the same way as the other adsorbed film emitters, i.e. by maintaining a balance between loss of barium and its replacement by outward diffusion.

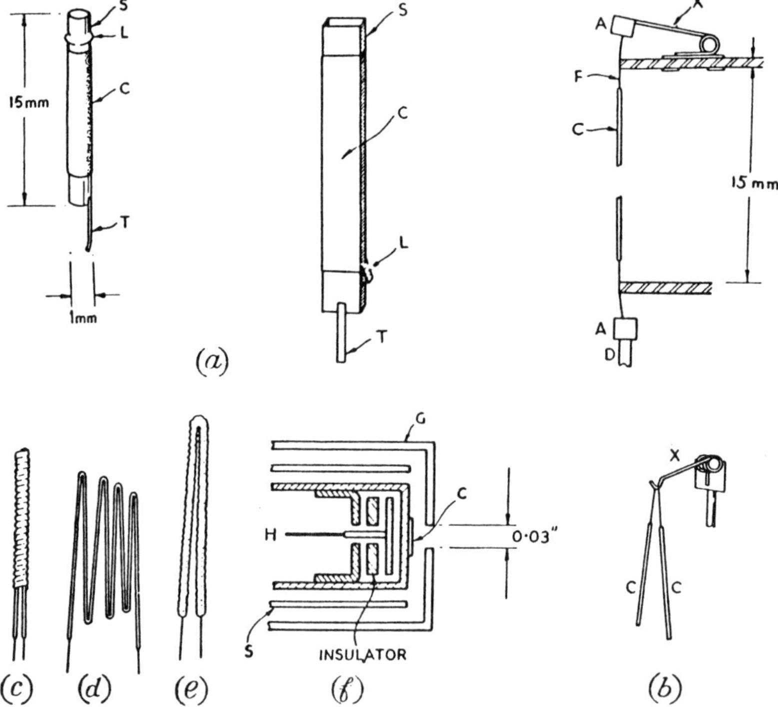

Fig. 1.3 Oxide-coated cathodes and heaters. (a) Receiving-tube cathodes. S—nickel sleeve. L—locating shoulder or tag. T—tail. C—oxide coating. (b) Receiving tube filaments. F—nickel or tungsten filament. C—coating. A—tabs. D—connecting lead. X—tensioning spring. (c) Alumina-coated double or single helix heater. (d) Alumina-coated faggot heater. (e) Alumina-coated folded spiral heater. (f) Cathode-ray tube cathode. C—coating. G—grid. S—heat shield. H—heater.

Fig 1.3 shows the arrangement of oxide cathodes typical of those in common use. They are prepared by applying a coating of the carbonates in an organic binder to a nickel cylinder or other suitable base or to a tungsten or nickel filament. The binder is usually collodion in amyl acetate in which the carbonates are held as a suspension, the particle size ranging from 1 to 10 microns. The coating is applied by painting it on, by spraying, or by drawing the filament through a trough containing the suspension and then through a drying oven. For receiving valves the coating is applied to a thickness of about 50 microns in a few layers, each being allowed to dry for a short time before the next is applied. Where an extremely even surface is required, as in a close-spaced planar triode, a finer powder is applied in as many as ten layers to a thickness of 10 microns. Thicknesses up to 100 microns may be used in larger valves. After spraying or painting, the cathodes are baked at $200°$ C to harden the coating and remove traces of moisture.

Activation of the cathode takes place in two stages, thermal dissociation of the carbonates at about $1200°$ K, and dissociation or reduction of the barium oxide to provide the excess barium essential for emission. The second process may be electrolytic or chemical, depending upon the choice of base material. If the core metal is pure nickel and does not react with barium oxide at the maximum permissible processing temperature, the cathode is activated at about $1200°$ K by drawing increasing current from the cathode with a suitable voltage applied to the other electrodes through a current-limiting resistor. The mechanism is that oxygen ions migrate outwards through the crystal lattice under the influence of the electric field in the coating, escape and are removed, while the barium ions move inwards to the core, from which they diffuse outwards after gaining an electron. They too will be lost by evaporation if their surface concentration is excessive. Emission is maintained at the correct operating temperature, i.e. below $1100°$ K, provided the rate of loss of barium is small. This requires very low gas pressure, for barium can be lost by combination with ions at the surface. Reduction of barium oxide is the only way to maintain the necessary surface concentration of free barium when the gas pressure is in the range 10^{-5} to 10^{-6} mm Hg, which is typical of receiving tubes, and so most of the cathode core materials used in quantity production are chemically active. Tungsten filaments themselves provide an adequate reducing action for the maintenance of emission for long periods at the pressures obtained by automatic pumping machinery. Nickel cores contain small percentages of a suitable additive such as silicon or tungsten. Activation then takes place by reduction of the oxide at about $1350°$ K. In some cases, particularly when the pumping time is short and the final pressure higher than that obtained after firing the getter, the process is completed by electrolysis in the ageing equipment.

While electrolysis of the coating provides a mechanism for activation and for replacement of surface barium, it can reduce life considerably by causing excessive concentration of barium at the surface if the ion current density exceeds a critical value. Cathode activity must therefore be high if long life is to be obtained, for this increases the ratio of electron to ion current in the coating. This requires that impurities which impair the semi-conducting properties of the oxide are removed during the exhaust process, and that others are prevented from entering it and destroying activity during life. It is for this reason that high outgassing temperatures are essential to successful operation for very long periods at current densities above $100\,mA/cm^2$, which is the maximum permissible in glass-envelope valves baked at temperatures no higher than $450°$ C.

Apart from these phenomena, a limitation on the maximum d.c. emission is set by the I^2R loss in the coating, this being further reduced by the possibility that local variations in the contact between coating and core will produce incandescent spots. At $1000°$ K, the permissible maximum continuous current is about $500\,mA/cm^2$ if high temperatures have not been used in processing. This is slightly higher than the maximum allowed by the electrolytic effect alone.

The work function of an activated oxide cathode is about 1.5 V, roughly equal to that of the best monolayer cathodes at their higher operating temperatures. The oxide-coated cathode is thus potentially the most efficient and long lived of all thermionic emitters. The saturated emission available continuously at $1000°$ K is theoretically $14\,A/cm^2$. In practice is is found to be no more than $1\,A/cm^2$, and under normal conditions of operation is limited to about $500\,mA/cm^2$. Even so, with a thermal radiation loss of $2\,W/cm^2$, the oxide cathode is still about three times as efficient as thoriated tungsten. It can therefore be used economically in valves where the power output is in the order of watts with anode voltages below $500\,V$. For example, a 6-W cathode supplies the emission in a beam-tetrode amplifier giving $20\,W$ output at $400\,V$, and $1\,W$ cathodes are used almost universally in subminiature valves where the anode dissipation is about $2\,W$. In both cases, a considerable fraction of the input power is lost by conduction or by radiation from areas not covered by the emitting surface. In directly heated valves, the efficiency is higher, and $10\,mA$ space-charge limited emission can be obtained with a filament input power of $100\,mW$. Oxide cathodes are therefore used universally in low level amplifiers and in power valves with outputs up to a few hundred watts, where the average space-charge limited emission is below $200\,mA/cm^2$. By reducing the operating current density and ensuring that loss of barium occurs only by evaporation, lives of many tens of thousands of hours may be obtained.

Oxide-coated cathodes also have the property that very high currents can be drawn for times shorter than a few μsecs. In a typical case $100\,A/cm^2$ is obtained at $1100°$ K with pulse lengths below $1\,\mu$sec. At 1 millisecond the emission is the same as that available continuously. It is this availability of high emission for times up to a few microseconds which makes oxide cathodes the only type which can be used in valves for the amplification of very short pulses. Many explanations have been offered for the difference in behaviour with time. None is entirely satisfactory. It can be attributed to removal during the pulse of electrons from the body of the material at a rate too large for them to be replaced by electrons crossing the interface, or to migration of barium away from the surface or of oxygen to the surface under electrolysis arising from the high current density. A rise in interface resistance would produce a similar effect. So would poisoning of the emitter by gaseous ions drawn into it by the d.c. field.

Heaters

Many electron devices require a unipotential cathode for their correct opera-

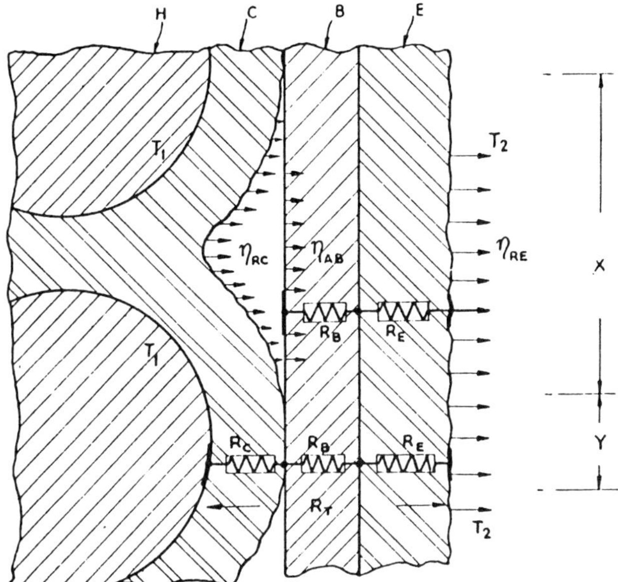

Heat transfer mechanism. *H*—heater at temperature T_1. *C*—heater coating. *B*—cathode base metal. *E*—emitter with surface temperature T_2. η—thermal radiation or absorption efficiency. *R*—thermal resistance. *X*—transfer by radiation, absorption and conduction. *Y*—transfer by conduction.

Fig. 1.4

tion. In most circuits, it is an advantage to operate the cathode at other than earth potential. Moreover, the direct-heating current of a nickel-based cathode similar to those in Fig 1.3 (*a*) would be orders greater than that which the seals in receiving valves could carry. For these and other reasons an indirectly-heated cathode is preferred in all valves other than high power amplifiers, except in the special case of lightweight portable equipment where economy in heating power is of first importance. The heater in an indirectly heated cathode supplies energy to the emitter by conduction, but a considerable fraction of the total is transferred by radiation and absorption. The transfer mechanism may therefore be represented by Fig 1.4. No detailed analysis is required to show that even if it is assumed that the total area of the heater is available for radiation, it must operate at a considerably higher temperature than the cathode surface. In practice, the effective area of the heater is only a fraction of that of the emissive surface, and there is considerable loss of heat by end cooling of the cathode core. Thus the heater in a typical oxide cathode with a surface temperature of 1000° K operates at between 1300° and 1500° K. The use of barium oxide in indirectly-heated cathodes is therefore essential, for any other emitter would require the heater to operate above the safe working temperature of any known heater or insulator material. Tungsten and molybdenum-tungsten alloy are the preferred heater materials because of their high tensile strength and low vapour pressure.

If full advantage is to be had from the ability to operate the cathode at other than earth potential, the insulation between cathode and heater must have high resistance and high dielectric strength. Pure alumina coatings are therefore applied by spraying or electrophoresis to a thickness up to 200 microns, the particle size being about 15 microns. After drying, the coated heaters are sintered at about 1900° K in hydrogen to form an adherent skin. The heater-cathode voltage can be as high as 250 V and the leakage resistance tens of megohms after an operating life of thousands of hours, but electrolysis will eventually cause breakdown of the insulation in receiving tubes if the heater is held more than 40 V positive to the cathode.

Secondary emitters

In high power magnetrons, primary emission is the means for starting oscillation, but secondary emission provides the major part of the operating current as a result of back-bombardment of the cathode by electrons accelerated by the r.f. field. A sprayed coating of barium oxide gives a secondary emission ratio (δ) between 2 and 2.4 at room temperature for either pulsed or d.c. bombardment with electrons of energy varying from 275 to above 800 V. The variation of δ with temperature is small up to 1000 K. Because of this and their high reserve of primary emission, oxide-coated cathodes have advantage over all others in pulsed magnetrons, provided limitations due to their lower electrical and thermal conductivity do not intervene.

Secondary emission is almost always a deleterious effect in space-charge control valves. For example, deposition of barium or its oxide changes the contact potential between the grid and cathode, and so modifies the grid control characteristics in the region of high trans-conductance. It also alters the grid-current characteristic by increasing either its primary or secondary emission, and increases noise level, particularly when the valve is fed by a high impedance source of signal. Secondary emission from insulators or the glass envelope produces noise or parasitic oscillation by leakage effects and modification of the electrostatic field near the electrodes. Emission from grids is inhibited by gold plating or coating them with a thin dense coating of finely divided platinum and carbon sintered on. Emission from insulators is minimised by shielding them from the electron stream, and the effects of electron bombardment of the glass by an internal coating of lamp-black or a metal cylinder enclosing the electrode system.

2 PHOSPHORS

Introductory

Phosphors are materials which will absorb energy and release it at a rate higher than that due to black-body radiation at their own temperature, the emission spectrum of the radiation being characteristic of each different material. Emission during the period up to 10^{-8} seconds after the excitation begins is termed fluorescence, that following and persisting after the excitation is removed, phosphorescence. The two together are called luminescence. In another terminology, fluorescence describes emission during excitation and phosphorescence the afterglow. The phosphors used in electron devices are cathodoluminescent materials which absorb the energy of incident electrons and release it as radiant energy in the spectrum range from infra-red to ultra-violet. Those most widely used give visible output, so permitting direct observation of waveforms applied to the deflector system of a cathode-ray tube. By density modulation of the electron beam and its concurrent deflection in two perpendicular directions to form a raster of varying intensity, a visible image may be reproduced. This is the familiar television picture. For observation of waveforms a green phosphor is preferred, because it matches the peak sensitivity of the eye and so produces less ocular fatigue. Where it is desired to record them by photography, a blue phosphor is used for its higher actinic value. In radar systems, phosphors are selected for their long afterglow, which gives an image adequately free from flicker in spite of the long interval between successive excitations which is fixed by the speed of rotation of the aerial system.

A phosphor consists of a *host* crystal, preferably colourless and having a high melting-point, which contains a suitable amount of an *activator* impurity. Host crystals should have high symmetry. Those used most often are cubic, hexagonal, rhombohedral or tetragonal systems of singly valent elements, such as Be, Mg, Ca, Zn, Cd, Al, Si, Mo or W, combined with O, F, S, or Se. The activator may be formed by selective decomposition of the host to produce a system such as ZnO (Zn), ZnS (Zn) or $CaWO_4$ (W). Alternatively it can be introduced as a controlled small amount of one or more multivalent elements, e.g. Cu, Ag, Mn, Pb, the quantity varying according to the desired output spectrum and/or the peak level of excitation.

Preparation and deposition of screens

Changes in impurity content below 1 part in 10^5 can produce great changes in emissive properties of a phosphor. It is therefore necessary to use starting materials of impurity content less than 1 part in 10^6, compared with the "reagent" standard of 1 in 10^4 and the "spectroscopically pure" 1 in 10^5, and to ensure that there is no contamination by the containers or atmosphere used in producing the crystalline material. The mechanical condition of the crystals likewise has a profound effect, and means must be provided for fine control of annealing cycles and furnace temperatures ($\pm 10°$ in $1500°$ C). After crystallisation, the material is reduced to particles between 1 and 10 microns in diameter, and special precautions must be taken to minimise the effects of rough treatment during the ball-milling process used for this purpose.

The powdered phosphor may be applied to its bearing surface in an organic binder which is removed during the bake, but in most cathode-ray tubes it is either *dusted in* or *settled*. In the first method, phosphoric acid vapour is blown into the bulb to clean the surface to be coated, and followed by a flush of clean dry air. The phosphor is then blown in as a suspension which adheres to the area attacked by the vapour, any excess being removed by tapping or shaking the bulb. This process is used for small-quantity manufacture of tubes for use in oscilloscopes or where settling in water is not practicable because of chemical action with the phosphor. Screens in television and in radar PPIs, which require a very small variation in coating thickness from tube to tube and over the face are prepared by the second method. The exact quantity of phosphor required to produce a coating of the correct thickness is mixed with demineralised or distilled water containing a small amount of an electrolyte such as ammonium carbonate. This prevents uneven settlement of the powder due to acquision of charge by the phosphor because of the different dielectric constant of the two materials. The well-shaken mixture is poured into previously cleaned and washed bulbs and allowed to stand for several hours, being kept free from vibration, draughts and other disturbances which would produce local variations in coating thickness. When the electrolyte is clear, it is decanted by slowly tilting the bulbs so that the settled layer is not disturbed. Each bulb is then dried by a stream of warm air, stoppered, and stored until the tube is completed by deposition of the graphite wall-anode or the aluminium screen backing, and sealing in of the gun assembly.

Emission spectra

The activator may either intensify or evoke a spectrum characteristic of the

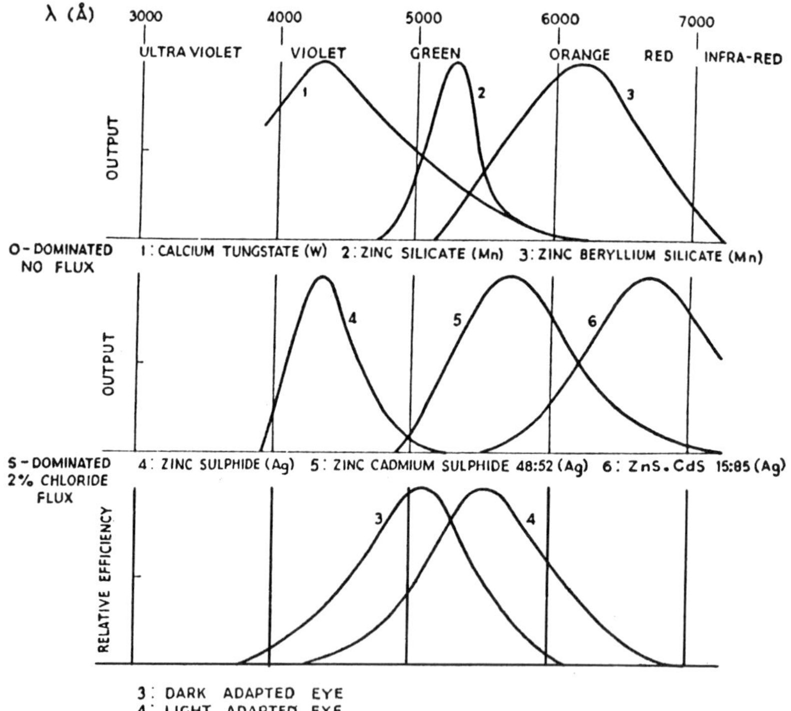

Spectral response of some commonly used phosphors
Note.—Peak phosphor output differs according to chemical composition, scale is in arbitrary units for each.

Fig. 2.1

host material, as for example 0.01% Ag in ZnS.CdS, or originate a spectrum characteristic of itself and suppress or diminish that of the host, as occurs with 1% Mn in ZnS or Zn_2SiO_4. In either case, the wave-length of the emission peak is increased. Other impurities may be added to inhibit radiation at some particular wave-length or modify the decay characteristic, as does Be in Zn_2SiO_4 (Mn). Thus, by choice of the correct combination of hosts, activators, inhibitors and heat treatment, either deliberately or accidentally, a phosphor may be prepared with almost any characteristics. This accounts in some measure for the wide variation in properties of phosphors having nominally similar ingredients and processing.

Fig 2.1 shows the emission spectra of some commonly used phosphors, together with the variation in sensitivity of the human eye. It will be seen that the response of manganese-activated zinc orthosilicate (synthetic "willemite") matches that of the dark-adapted eye and is easily detected by the light-adapted eye. For this reason, mineral willemite was the obvious choice as a screen material in oscilloscope tubes for many years before a reliable synthesising process had been evolved.

Efficiency and overload characteristics

Phosphors are energy converters, and so power that is not radiated will raise their temperature by an amount inversely dependent on their efficiency. As the temperature rises, efficiency decreases slowly until a critical temprature is reached in the range 300–700° K, when it falls to zero within about 100°. The efficiency of a phosphor rarely exceeds 10%, and so the maximum permissible mean power input is that which would raise the screen up to the critical temperature. The peak input power is determined by saturation effects in the luminescent material. At any fixed voltage, output increases with beam current until the phosphor penetrated by electrons is unable to convert any more energy into radiation. An increase in output can then be obtained only by increasing the voltage and so the depth of penetration of the beam until internal heating and absorption and scattering of radiation by the phosphor begin to offset the potential gain, or until electrons pass through the crystals without giving up their energy. The particle size and screen thickness must therefore be correctly chosen to obtain the maximum conversion efficiency and peak and mean output from an electron beam at any particular voltage.

For a phosphor to have any practical value, it must give sufficient output to be viewed in comfort by an observer. It must therefore produce a certain minimum amount of radiation per unit area before saturating. This amount increases with the ambient light intensity. Where waveforms are to be observed it is permissible to operate near the saturation level, and so phosphors having high output at low voltage are preferred, for this simplifies problems of focusing and deflection of the beam. For presentation of radar information, the saturation level determines the highlight intensity and so the maximum contrast that can be obtained in the presence of ambient light. The operating voltage is therefore higher than that of instrument tubes, and the phosphor selected for its maintenance of efficiency at high levels of excitation.

Electrical properties of phosphors

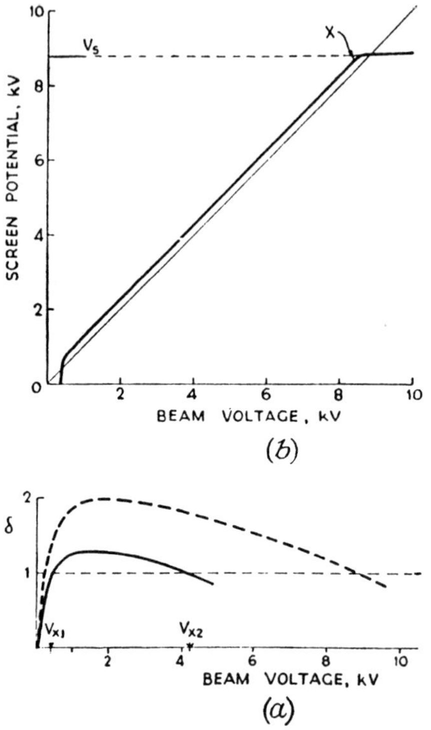

Electrical properties of phosphors. (a) Secondary emission ratio.
(b) Screen potential *v.* beam voltage.
X—break point. V_s—striking potential.

Fig. 2.2

Phosphors are either insulators or poor semi-conductors. The operating voltage of the screen therefore depends upon their secondary emission characteristics, which have the general form of Fig 2.2 (*a*). When the beam voltage is below the first crossover value V_{X1} the screen is charged negatively and reflects the incident electron beam, thus giving no output. Between the first and second crossover, the screen potential is very close to the beam voltage, and so the phosphor can be deposited on an insulator selected for its good optical transmission properties. Above the second crossover V_{x2} the screen potential stabilises at a value close to V_{x2} irrespective of the beam voltage. The relation between beam voltage and screen potential is therefore as shown in Fig 2.2 (*b*), where V_s is known as the *sticking potential*. The value of V_s may be between 3 and 50 kV, depending on which phosphor is used, what binder is incorporated, and how it is deposited. For operation with beam voltages above 10 kV, the sticking effect is overcome by depositing a layer of aluminium about 0.1 micron thick on the bombarded side of the screen and connecting this to the final anode. The aluminium also acts as a reflector of visible output, so almost doubling efficiency at any particular energising voltage. The metal backing also reduces the effect of ion bombardment of the screen, for a layer which is easily penetrated by the electron beam can be almost impervious to ions. The energising voltage V_E is less than that of the beam V_B by an amount V_A representing the thickness of the backing. The effective energising voltage with an unbacked screen is also less than the beam voltage because of the effects of secondary emission, but this is a smaller factor than V_A. Thus as the beam voltage is increased from zero the unbacked screen has higher efficiency until the *break-even* value is reached. Above this the aluminised screen has higher efficiency. The break-even voltage for sulphide screens is between 5 and 10 kV, that for the silicates somewhat higher.

3 MECHANICAL DESIGN

Introductory

Successful construction of an electron device depends upon creation of the correct environment for the moving electron. This should remain unchanged throughout the operating life of the tube. For tubes to be of any practical use, their characteristics must not vary significantly from sample to sample. This requires their construction from material of constant properties by processes which are repeatable at will. Their mechanical design must be such that they have adequate strength for their intended use, and that distortion during manufacture or life does not produce variation in characteristics. Thus, in addition to selecting or evolving an electron-field interaction process having the desired electrical effect, the designer of an electron tube must adapt for his own purposes the best available techniques in chemistry, metallurgy, mechanical engineering and vacuum physics. The correct environment for the moving electron is either high vacuum or a gas or mixture of gases at a chosen fixed pressure, within which is placed an electrode assembly free from contaminants in the form of occluded impurities, surface films or other foreign matter. In other words, an electron device should comprise clean envelope, electrodes and insulators, an efficient source of electrons, and nothing else unless its operation depends upon inclusion of a phosphor, a gas or some other chosen substance. This condition should also be fulfilled if experiments to prove the feasibility of a proposed interaction process are to have any significance. If it is not, unpredicted and misleading effects will be observed. Techniques used in experiment or manufacture must therefore give the necessary freedom from impurities without detriment to the mechanical properties of the completed device.

Mechanical design

Electron devices must be fabricated from the minimum number of parts, for each is a potential source of contamination or inaccuracy. The parts must be joined so that there are no pockets between them from which gas will escape during and after the evacuation process. Where surfaces are in contact over a large area they must be brazed or glazed together for the same reason. Correct mechanical design is achieved when the total area of the surfaces exposed to the vacuum is reduced as far as possible. This implies that the surfaces themselves be smooth and free from fissures.

During operation of a thermionic device the cathode is at a temperature above 700° C. Other parts of the assembly are heated by radiation or collection of electrons to temperatures at which the rate of arrival of energy is balanced by the rate of loss of heat to their surroundings by conduction, convection or radiation. The tube must therefore be designed so that the cathode is thermally insulated from and at the same time accurately located with respect to the other electrodes. Conversely, the other parts of the assembly must be cooled efficiently so that unwanted electron emission is not obtained or impurities in the electrodes and insulators released by elevation of their temperature. The design must also accommodate the different thermal expansion of the various parts of the assembly by providing freedom for extension without detriment to the strength or electrical properties of the tube. Structural materials must have adequate hot strength, which may or may not be predicted from that at normal ambient temperatures, and be free from *creep* or distortion under stress or repeated cycling from low to high temperature. Glass or ceramic parts should be subjected to no great tensile or bending stress. They must be placed so that material volatilised from the hot parts of the assembly is not deposited on them, and be screened from electron bombardment, for this can impair the insulating properties or build up surface charges and so distort the field configuration in the interaction region.

Electron devices must be built to fine limits if they are to function in the predicted manner, and so materials used for fabrication of the critical parts of the assembly should be capable of accurate shaping by normal workshop and factory techniques. Alternatively, the design must be such that critical dimensions can be adjusted during assesmbly, and that after adjustment they remain fixed. Thus the mechanical problems in engineering an electron device are essentially the same as those in any other structure. In brief, they are to ensure that the assembly is adequately free from distortion under stress, stable with time and as easy as possible to fabricate. The techniques used are familiar in other contexts, except that they may be on a smaller scale and modified in detail to meet the special requirements of vacuum technology. For example, the cathode in a receiving valve is an end-supported beam. Three-point attachment is used to mount the cathode and grids of small transmitting tubes. Parts made of sheet metal are stiffened by ribs formed by pressing channels in the flat surfaces and joined by tabs or rolled-over seams.

Receiving or general-purpose tubes

Fig 3.1(a) is a simplified plan and section of a pentode used in communications and similar equipment. The insulators are about 0.01" thick, and made from first quality ruby mica, which has poor thermal conductivity and high insulation resistance. Holes are punched in them for location of the electrodes and to increase the leakage paths. Peripheral projections or *snubbers* hold the assembly or *cage* securely within the glass envelope. The cathode K is a nickel-alloy cylinder securely anchored in one mica but free to expand by sliding in the other.

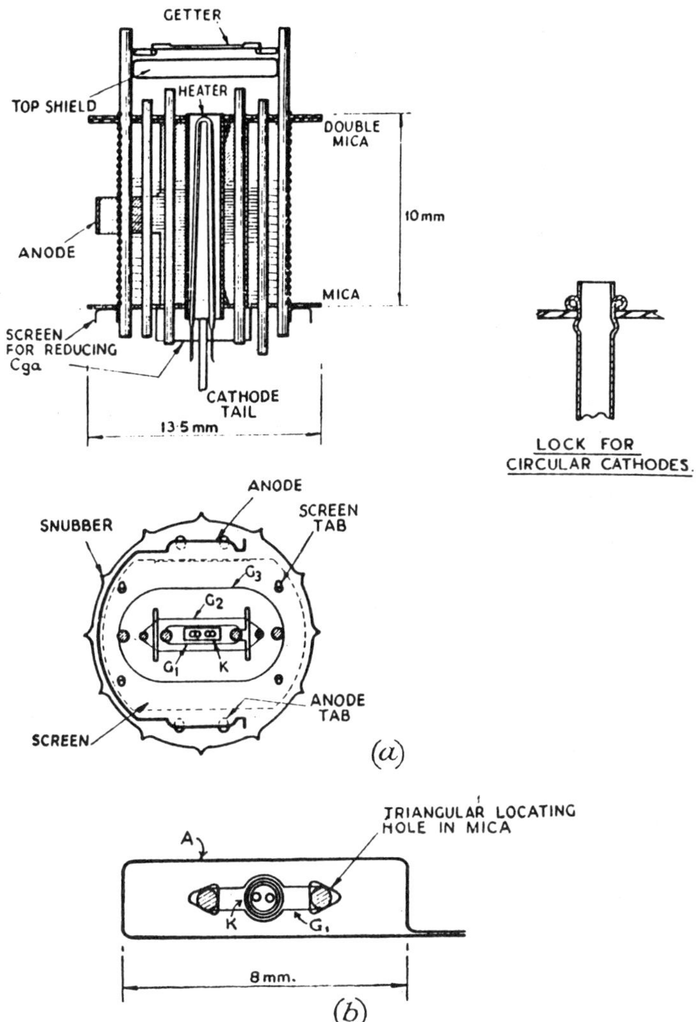

Receiver tubes. (*a*) Pentrode electrode system. Spacing: K to G_1 0·004 in., G_1 to G_2 0·018 in., G_2 to G_3 0·074 in., G_3 to A 0·1 in. Pitch: G_1 150 t.p.i., G_2 65 t.p.i., G_3 36 t.p.i. (*b*) Triode electrode system. Spacing: K to G_1 0·0025 in., G_1 to A 0·0125 in. Pitch: G_1 220 t.p.i.

Fig. 3.1

Connection is made to it by a compliant *tail*. It is coated with a thin uniform layer of the mixed oxides of barium and strontium. The heater H is a close-wound spiral or a folded filament of tungsten or molybdenum-tungsten alloy, and coated with alumina sintered on before insertion. It is held in position partly by friction, partly by the *legs* which are welded to straps securely fixed in the lower mica. The control grid G_1 is an elliptical spiral of fine molybdenum, manganese nickel or gold-plated tungsten wire wound on a pair of chrome-copper alloy *side-rods* which are a tight fit in their locating holes. The turns of the spiral are anchored to the side-rods by welding or gold brazing or by *nicking* and *swaging*. This is done by rolling a groove in the rods into which a turn is laid, and then rolling again so that metal pushed up in nicking is forced down to grip each lateral as it is wrapped round the support. Being good conductors, the side-rods keep the grid temperature uniform along its length. Where necessary, extra cooling is provided by flags mounted on their upper ends. The outer grids G_2 and G_3 are made and anchored in a similar way to G_1, except that nickel or manganese-nickel side-rods may be used. When high transconductance is required, the spacing of the grid wires and their separation from the cathode is about 0.05 mm, and their diameter around 10 microns. To achieve the necessary uniformity, stability and mechanical strength, it is then essential to use a *frame grid*, in which the side rods are joined by cross-members before the turns are wound under tension and gold brazed in position. When the valve is to be used in aircraft or other mobile equipment and so subjected to considerable mechanical shock and vibration, double micas are used at each end of the cage, the holes in one being slightly offset with respect to those in the other, and the side-rods welded to brackets previously anchored in the top mica. The carbonised nickel or aluminium-clad iron anode is the main structural member of the electrode assembly. Tabs at the end of the vertical portions hold the insulators in fixed location relative to each other, being turned down at right angles or having straps welded across them to grip the micas against the anode edge. Connection between the electrodes and the sealing wires in the base is by welded joints below the lower mica.

The critical dimensions of the assembly are measured in thousandths of an inch and the tolerances on dimensions in fractions of a thou. Thus each part must be accurately formed from material of controlled mechanical properties. Tubes of this kind are mass-produced, the piece-parts being made on automatic machinery and manually assembled. Each piece must therefore be strong enough for manipulation without distortion and yet as light as possible to minimise pumping time and the effect of vibration on the completed unit.

Fig 3.2 shows the construction of a planar triode used as a signal amplifier and oscillator at frequencies up to 3 mHz. The cathode is a thin dense coating of mixed oxides on a comparatively thick nickel disc attached to the cup enclosing the heater. The emitting surface is pressed or ground accurately flat and perpendicular to the axis of the assembly as the final operation in making the cathode sub-assembly. The gold-plated tungsten wire grids are wound under tension with up to 400 turns per cm across an annular molybdenum frame and gold brazed in position. The diameter of the wire is typically 6 microns (0.006 mm), and the active area of the grid about $0.2\,\mathrm{cm}^2$. It is essential that the grid wires be uniformly spaced over the active area and that none be slack or have nodules of gold adhering, for local variations in grid-cathode spacing will reduce the signal to noise ratio of the tube as a low-level signal amplifier. The body is assembled in jigs so that the anode and cathode mounting tubes are accurately aligned, and the grid made perpendicular to the axis by pressing it down and stretching the compliant section which joins it to the grid ring. The cathode sub-assembly is then inserted and the grid-cathode clearance adjusted by means of a microscope before the locking weld is made. The anode-grid spacing is set by measurement of the capacitance.

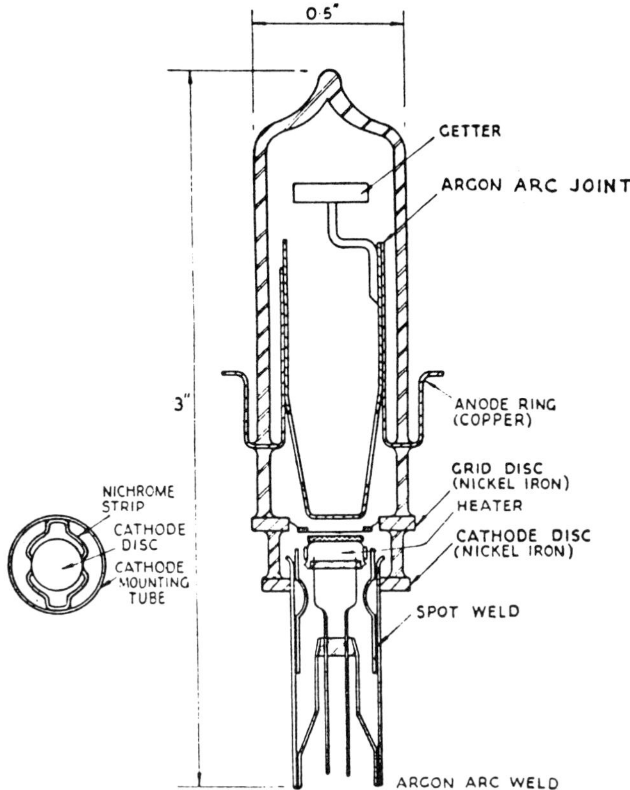

Fig. 3.2   Planar triode

Microwave tubes

So far we have considered the design of assemblies used at low frequency or at wavelengths large compared with their dimensions. At wavelengths below 30 cm, the designer is faced with problems in addition to those we have touched on, particularly when high power output is required. Irrespective of the interaction process, the active parts of the device are limited in size to dimensions comparable with one half-wavelength, and so the power input per unit volume may be very much greater than that in valves operating at lower frequencies. This poses problems in cooling, maintenance of high vacuum in the presence of high-intensity electron bombardment of internal surfaces, and in coupling out the r.f. power. Where a tuner is not fitted, the resonator system must be designed and built so that there is the minimum variation of frequency from sample to sample and during processing and operation of the valve. The ensure high efficiency, ohmic losses accentuated by the skin effect must be minimised by correct placing of the joints between the separate parts of the assembly and by making all surfaces as smooth as possible. In magnetrons, a compromise must be accepted between over-heating of the cathode by back-bombardment and the power required initially to bring it to the correct operating temperature. Engineering microwave tubes is thus a very complex technology. We shall consider here only a few typical devices which illustrate how some of the problems have been solved.

Magnetrons

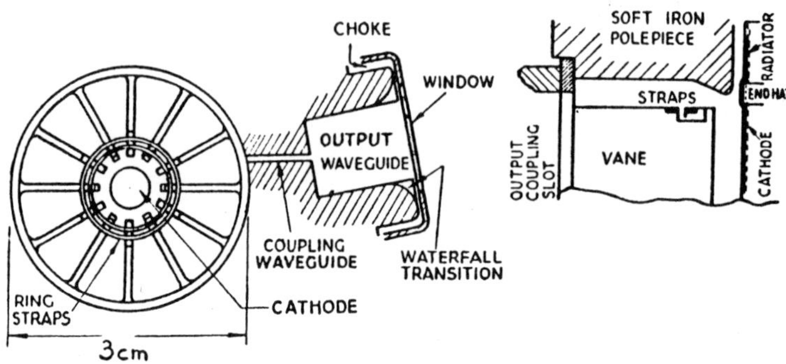

Fig. 3.3

Fig 3.3 shows the cross-section of a 3-cm magnetron giving 200 kW peak 200 W mean output. To prevent deposition of volatile material on the window during processing and operation of the tube, the output coupling section is set at an angle so that the window cannot see the hot cathode. In some arrangements, the coupling transformer may taper exponentially up to the narrower dimension of the output waveguide. The resonant frequency of the cavities must be kept within very close limits, and so the method of fabrication must give high dimensional accuracy without requiring long and costly processes. Anodes can be machined from the solid, but this is expensive and difficult with O.F.H.C. copper, becoming more so as the number of segments increases. Tubes operating at 3 cm and above are therefore made by brazing a set of accurately shaped vanes into a cylindrical shell. At and below 3 cm the necessary accuracy can be obtained by *hobbing* a plug of vacuum-cast copper. The hob is ground from hardened steel so that its cross-section is that of the hole in the anode. The plug is placed in a close-fitting steel cup with a small-pressure relieving hole in the bottom and heated to about 400° C, when the hob is forced into it by a hydraulic press. After cooling, the hob with the plug attached is withdrawn from the cup and the outside of the copper turned true with the hob axis. After the hob is removed, the hole in the copper is filled with a polymerising resin and the ends of the plug turned to the required shape, including the recesses for the straps. The anode is then chemically cleaned, and polished by deplating in a suitable electrolyte before the other parts of the assembly are attached. Hobbing is also used for production of 'rising sun' anode blocks.

Magnetrons differ from all other microwave devices in having the cathode at the centre of the resonator system. Its position therefore has a significant effect on performance. Thus all materials used in its construction must have mechanical stability and strength at high temperatures if the predicted characteristics are to be obtained and remain unimpaired during life. They must also be non-magnetic, or nearly so (thin nickel cylinders carring the cathode coating are acceptable). Side-arm cathode supports are rigid tungsten rods.

Other devices

Although almost all electron devices require some mechanical design techniques which are applicable to them alone, they all have some features in common. For example, cathode-ray tubes have details in common with receiving valves, but must be designed to avoid implosion under atmospheric pressure on the face. Because a small number of electrons is sufficient to produce visible spots, field emission must be prevented in all parts of the assembly where the operating fields could accelerate them to the screen. Surface charges on the insulators can distort the fields, and so the electrodes are supported by lugs anchored to two or more longitudinal glass or ceramic rods which are shielded from electron bombardment by the anodes and deflector plates themselves. The electron-guns in reflex klystrons and cathode-ray tubes use basically similar geometry but differ in detail because of the higher operating current of the microwave tubes.

4 TECHNIQUES OF CONSTRUCTION

Introductory

All materials used in fabricating electron devices must be easily cleansed of unwanted inclusions, and have low vapour pressure at the highest temperatures to which they are subjected during processing and operation of the completed tube. They should be chemically stable and inert, or at least non-reactive with materials with which they are in contact. There are exceptions to this, e.g. brazing materials which alloy with metals they join, and cathode-base materials which have a reducing action on the oxide coating. They must also be mechanically stable to ensure that the required accuracy of construction is achieved in the finished tube and maintained during its life.

Electrodes

Metals used in valve construction must be suitable for accurate shaping by machining, pressing or folding. They should therefore be either stiff and free-cutting or ductile and malleable in the working state. It is an advantage if they have high mechanical damping factors to minimise the effect of vibration in use. They must have adequate strength at high temperatures and be free from creep or recrystallisation under temperature cycling during processing and operation of the valve. Their coefficients of thermal expansion should either be low, or matched to those of the materials to which they are joined. Thermal conductivity and emissivity may be either high or low depending upon the intended use. For example, cathode assemblies have low-conductivity materials for mounting the active element and highly reflective surfaces on shields to minimise heat loss, but anodes are made from material having high emissivity and/or conductivity. The melting-point must be well above the highest temperature reached in processing. Electrical conductivity should in general be high, but moderate conductivity is an advantage where spot welding or r.f. induction heating is used for joining piece-parts. Primary and secondary emission should be low unless the metal is used as a cathode. Resistance to oxidation is an advantage, although the increasing use of vacuum and hydrogen furnaces for pretreatment and assembly processes makes this less important than in the past. Most important of all, metals should absorb only a small amount of gas and give this up readily when heated in vacuum. Exceptions to this are *getters* such as zirconium or tantalum, which are used to absorb gas evolved from other parts of the assembly and so maintain a high vacuum within the envelope.

No one metal has all these desirable properties. A compromise has to be sought in the choice of which shall be used for construction of a particular piece-part. Sometimes it is achieved by selecting a compound such as the nickel-iron and nickel-iron-cobalt alloys used for sealing to glass.

Nickel

Nickel is used in the greater part of the electrode assembly in low-power valves for radio, television and communications equipment. It forms the base for the cathode coating in these and many other tubes. Being easily stretched and showing no sharp break at the yield point, it may be pressed, rolled and folded without difficulty, and is readily joined by spot welding to itself and other metals. As normally supplied, it contains a small proportion of additives such as silicon or manganese which give it adequate hot strength and hardness for most uses. Where these additives have undesirable chemical properties, the electrolytically pure metal is used, and where greater hot strength is required, an alloy of nickel with tungsten. The latter is finding increased application as a cathode-base material, when it includes a small percentage of magnesium, aluminium or other reducing agent. Highly polished nickel may have an emissivity as low as 10% of that of a black body. Carbonising increases this to above 80%. Being slightly magnetic and having relatively high resistivity, nickel is easily raised to red heat for out-gassing by induced r.f. eddy currents, a method which is used universally in the manufacture of glass-envelope valves.

Molybdenum

Molybdenum is used in place of nickel where operating temperatures are higher or greater hot strength is required, e.g. in the centre support rod of a high-power magnetron cathode or the anode of a radiation-cooled transmitter valve. If adequately purified and free from oxide, it may be turned, pressed, folded or rolled into shape by proper design of tools and annealing processes, but is in this respect much inferior to nickel. It can be spot welded to nickel or iron without difficulty. Welds between molybdenum parts are more difficult to make because the metal is easily oxidised at 550°C. Thus the weld is brittle and tends to disrupt due to the formation of hot spots during the current pulse. This can be minimised by sandwiching a layer of nickel between the parts to be joined.

Iron

Iron or steel is used as a cheaper replacement for nickel in receiving-valve anodes, and for cathode supports in microwave tubes. Like nickel, it is easily shaped and welded and heated by induced eddy currents. Steel anodes are coated on the outer face with aluminium to increase emissivity. This produces a strongly adherent matt-black surface when the parts are raised to red heat during the exhaust process. Nickel-plating carbonised in the same way as nickel is also used for the same purpose.

Tantalum

Because of its high melting-point and low vapour pressure at temperatures up to 2,000°C, tantalum is used for the anodes of small high-power valves cooled by radiation through the glass envelope. It is completely outgassed by heating to above 2,000°C. At temperatures between 900°C and 1,100°C it will act as a getter, particularly for hydrogen, and is usually operated in this range to assist in maintaining a high vacuum. Its gettering properties and ease of heating by induced eddy currents makes it particularly useful for the construction of ovens in which to process other materials in an evacuated enclosure. It may be pressed and rolled into shape, but is expensive to produce and prepare in a form suitable for the construction of large electrodes, and so is used mainly for cylindrical or cup-shaped parts with diameters less than 3″.

Tungsten

Tungsten is used where high operating temperature, low vapour pressure and great strength are required together, as they are in filaments, cathode heaters and grid wires. Being extremely difficult to manipulate except by swaging, grinding or drawing, it is used mainly as rod or wire, which can be made with diameters down to 10 microns and chemically etched to 3 microns. The swaging and early drawing processes may produce longitudinal hair-line cracks in the surface of the metal. Tungsten rod used for making vacuum seals is therefore ground to size before glazing. The metal is prone to embrittlement due to crystal growth at high temperatures, and small amounts of inhibitors such as thorium may be added to material intended for use under tensile stress. Where a material more readily worked is required, an alloy of tungsten with molybdenum is used, which also is less liable to crystal growth than the pure metal. The alloy cannot be used as an oxide cathode base material.

Copper

Copper is used where high electrical and thermal conductivity is of first importance. It therefore has its widest application in the construction of high power transmitting valves with external anodes cooled by water or air blast, and in microwave tubes where the resonant elements form part of the active volume. Alloyed with a small percentage of chromium to increase its rigidity, it is also used for the grid side-rods in receiving and small transmitting valves. Thus the grid is maintained at as even and as low a temperature as possible, and can if necessary be cooled by radiation from flags mounted on the support rods clear of the hot region. Copper is easily pressed, folded or spun into intricate shapes, and can be machined to fine limits by the use of high cutting speeds and suitable tools and lubrication. It cannot be joined by spot welding, but is easily joined by brazing in hydrogen or by arc welding in an inert atmosphere. Copper is easily oxidised at temperatures above a few hundred degrees C. Moreover, cuprous oxide has a high vapour pressure. The temperature must therefore be kept below 1,000°C at all times during the processing of valves containing copper, and pumping speeds be high enough to remove volatile material before it is deposited with detrimental effect on the other parts of the assembly. All copper used in valve construction must be free of oxygen. If it is not, it becomes porous or mechanically weak the first time it is heated to red heat in vacuum or a reducing atmosphere.

Other metals

Although the metals referred to above meet most of the designer's requirements, other metals are used for special purposes in the construction of electron devices. Gold plating is used to inhibit emission from grids. Aluminium is used as the backing for phosphors in high-voltage cathode-ray tubes. But irrespective of which metals are chosen for a particular purpose, they must be supplied chemically pure and consistent in their other characteristics, for on this depends the successful completion of the device built with them. On his part the designer must ensure that no metal is subjected to conditions to which its physical and chemical properties are unsuited.

Insulators

Insulators used in valve construction must have adequate mechanical strength and good dielectric properties at their operating temperature. When they are used for electrode location it is an advantage if they can be fabricated with the necessary accuracy and the electrodes anchored in them by simple processes. Properly used, they are subjected to the minimum tensile stress, but the high tensile strength of mica and alumina is an advantage during assembly. They should not react with electrode materials, nor decompose at the temperatures during processing and operation of the completed device.

Mica

Mica is used for the electrode supports in nearly all receiving and other low power valves. Because of its crystalline structure it is easily split into thin laminæ, the thickness of those in receiving valves being about 0·2 mm. The laminæ are resilient and adequately robust for the manual assembly processes used in valve manufacture, and can be punched with the required accuracy or drilled by clamping between metal plates. Mica is friable, particularly at the edges, and if vibration is not to cause failure, the mechanical design must minimise the possibility of chafing by electrode supports moving in their locating holes.

Mica has very high specific resistance (up to 5×10^{16} ohm-cm) and dielectric strength (up to 200 kV/mm). It is limited in its application only by decomposition, which begins at about 500°C. This is below the outgassing temperature of the metals used for electrodes, and so the assembly cannot be properly processed during exhaust without risk of damaging the mica. This is overcome to some extent by the use of a getter film deposited on the inside of the envelope, but it has been shown that a threehold increase in cathode activity is obtained by replacing mica with ceramic and raising the processing temperature above 650°C.

Glass

Unless it forms part of the vacuum envelope, glass is rarely used as an electrode support because of its fragility. It can be used only where processing temperatures are very little higher than with mica, and must be incorporated as part of a sub-assembly where previously sealed-in anchoring lugs are welded to the electrodes, as in cathode ray tubes.

Ceramics

Ceramic insulators are preferred to all others when the problems of accurate shaping and assembly into electrode systems do not preclude their use on economic grounds. They thus have their widest application in power amplifiers, where their refractory properties can be used to great advantage, and sufficiently accurate location of electrodes can be obtained by processes which may not be amenable to mass production on the scale required for receiving valve manufacture. Pure alumina is used for heater-cathode insulation in indirectly heated valves, being applied to the heater as a suspension in an organic binder and sintered at about 1,600°C to form an adherent insulating layer.

Pure alumina bodies are difficult to prepare and machine, and so electrode supports are usually made from material containing a few per cent of flux, e.g. magnesia and sodium metasilicate, which reduces the firing temperature and makes them easier to shape. Even so, location of electrodes should be made with respect to a surface which can be machined by a comparatively simple operation, e.g. grinding one face or an outer cylindrical surface. Where this is not possible, anchoring lugs must be firmly attached to the insulators and correct alignment obtained by adjustment of the position of the free ends to which the electrodes will be anchored. Where easier fabrication is required and the mechanical stresses are such that the greater strength of alumina is not required, frequentite and steatite can be used to advantage, being much more easily machined before the final firing. All three ceramics may be raised without adverse effect to the outgassing temperatures of the metals commonly used in valve construction. They can be cleansed by firing in air at around 1,000°C.

Joining Processes

When joining electrodes and insulators, it is necessary only to obtain a good mechanical connection, and so the method is chosen entirely on that basis. In mica, which is resilient, electrodes are located and held by support wires or tabs push-fitted into holes or welded to lugs previously anchored in position. Similar techniques are used with ceramics, but because these are rigid bodies the electrodes are held by attaching them to anchor lugs or by distorting their supports to take up the slackness in fit between them and the holes in the insulator. Connections between metal parts must be both mechanically and electrically reliable. If they are not, connection to an electrode may be broken or electrical noise generated by the passage of current through a contact of randomly varying resistance. Failure of the assembly under mechanical vibration or shock will also occur. Piece-parts for microwave tubes are joined as a sub-assembly by brazing with copper or copper eutectic solders in a reducing atmosphere, but most of the joints in electrode assemblies for lower frequencies are made by electric welding, which eliminates the brazing material. Arc welding in a rare-gas atmosphere is used for joining low-resistivity material, but for most other purposes *spot welding* is preferred.

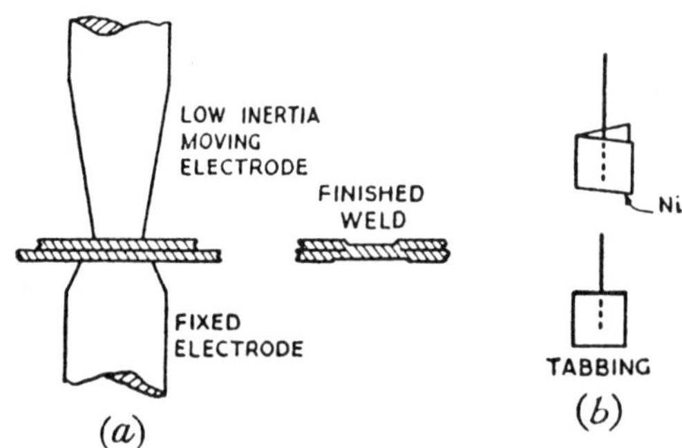

Spot welding. (a) Electrode system.
(b) Tabbing of fine wire.

Fig. 4.1

In this process (see Fig. 4.1) the parts to be joined are gripped between chrome-copper electrodes through which a controlled high current passes for a definite time. Intense local heating and deformation of the metal under pressure then produces a strong mechanical bond by fusion at the interface. At the instant of welding the hot metal tends to collapse and so lose contact with the electrodes. The upper electrode is therefore made as light as possible without detriment to its strength or current-carrying capacity, and spring-loaded or mounted on a pressure-actuated ram to obtain the proper follow-through as the weld is made. For reliable welding, the resistance between the electrode connections must be the same each time the operation is performed. The resistivity of the materials to be joined must therefore be closely controlled and their surfaces kept free from oxide or other contaminants. The electrode faces must be clean and parallel, and remain so during continued use or be faced-up at frequent intervals. The duration of the current pulse must be held to the predetermined value by means of a timer actuated when the pressure applied by the electrodes has reached a chosen value. Where the materials are prone to oxidation at the welding temperature, an inert gas atmosphere should be used. Protection can also be obtained by applying a few drops of methylated alcohol before bringing down the top electrode. Welds can be made repeatedly in this way between tungsten wires down to 0·001″ diameter, but for reliability and ease in assembly, fine wires are usually *tabbed* by placing them in the fold of a small rectangle of nickel, which is then welded together (see Fig. 4.1 (*b*)).

Vacuum Envelopes and Seals

In addition to having all the desirable properties of electrode and insulator materials, materials forming a vacuum envelope must be impervious to gases at its operating temperature. They must be sealed one to another by processes making vacuum-tight joints which are not impaired by subsequent processing and use of the completed device. The magnitude of this problem can be appreciated when it is realised that a microwave power valve may contain up to one hundred vacuum joints. Miniature valaves have eleven or more, and the design of mass-production methods to make these reliable is again a formidable problem. In many cases, the envelope is also the means for locating the electrodes, and then special care has to be taken to ensure it has the requisite strength and stability. Where the valves are to be used at high frequency, insulators forming part of the r.f. output circuit must have low dielectric loss, for this sets an upper limit to the power that can be coupled out from the interaction region. The materials used most often in vacuum envelopes are glass, copper, nickel-iron and nickel-iron cobalt alloys. For special purposes, tungsten, or molybdenum rods may be sealed into glass envelopes.

Glass

Glass forms part of the envelope of nearly all valves made at the present time, for it is cheap to produce and easily blown, moulded or pressed into almost any shape. Glass is a fused mixture of silica and metallic oxides, the silica content varying from 60 to 80% of the total weight. The characteristics are determined by the additives, and so many variations can be made that a glass can be found to suit almost any purpose. For example, the addition of lead oxide, sodium oxide and potassium oxide, or calcium oxide and sodium oxide makes the soft glasses used for receiving valve envelopes. The addition of boric oxide with small percentages of other oxides produces the borosilicate glasses used in higher power valves. Lithia borosilicate glass has very low dielectric loss at high frequency, and is therefore used for the r.f. output windows of microwave tubes. The range of melting-points is such that a glass can be made for sealing to mica and another for sealing to molybdenum. Being super-cooled liquids, glasses wet materials to which they can be sealed. This allows the seal to be made direct to a metal or a thin film of oxide on its surface, and increases the probability of producing reliable vacuum-tight joints. Glass has low tensile strength, and its coefficient of thermal expansion varies markedly with composition and temperature. Thus seals to metal and to other glasses can be made only by ensuring that no insupportable strain is set up during cooling of the joint. Glasses soften continuously when heated, showing no abrupt change in mechanical properties. The strain, annealing, softening and working temperatures are therefore defined arbitrarily in terms of viscosity. Glass must be worked at the lowest practicable temperature to minimise risk of devitrification, and should not be raised above the working point when making seals. If possible, it should be worked nearer the softening temperature and the seal made by the application of pressure normal to the joint. Strain in the glass can be relieved by maintaining it at a steady temperature between the softening and strain points and then slowly cooling to the lower temperature. Rapid cooling below the strain point produces no adverse effect. Glasses have two great disadvantages as envelope materials – their fragility and the limitation they set on operating and outgassing temperatures. Some also suffer from electrolysis at temperatures around 300°C.

Silica was used for the envelopes of radiation-cooled valves such as the high power transmitting triodes designed and made by Signal School from the 1920s. Its high softening point and thermal transparency permit operation without forced-air cooling at much higher power levels than with any glass. It cannot be sealed to any refractory metal, and so must be sealed to a series of glasses having gradually increasing coefficients of expansion before being joined to the molybdenum or tungsten electrode supports.

Glass Envelopes

All-glass envelopes are used for the majority of electron devices manufactured at the present time. Connections to the electrodes are made by lead wires sealed through a prefabricated part of the envelpe, the *button* or *stem*.

Fig. 4.2(*a*) shows two types of soft-glass envelope used in miniature receiving tubes. The leads comprise three parts butt-welded together, nickel for easy welding to the electrodes, copper-clad 42% nickel-iron alloy for the seal, and nickel for connection to the circuit in which the valve is operated. The alloy wire has radially a lower coefficient of expansion than soft glass and so produces compression at the seal. Strain parallel to the wire is relieved by flow of the copper sheath. The hard-glass envelopes of (*b*) are used for small transmitting valves. The leads are molybdenum or nickel-iron-cobalt alloy rods which carry the electrode assembly. The buttons are often prepared by placing pre-glazed pins in position

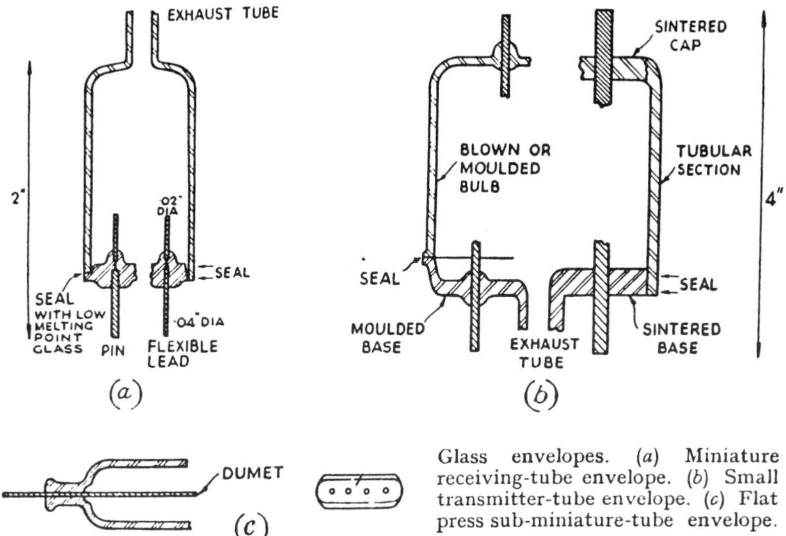

Glass envelopes. (a) Miniature receiving-tube envelope. (b) Small transmitter-tube envelope. (c) Flat press sub-miniature-tube envelope.

Fig. 4.2

in a mould, adding powdered glass and sintering at temperatures below the melting-point. There being less strain on the glass in a sintered assembly, bases can be made with larger and more irregularly shaped leads than could be housed in a moulded unit. The flat press of (c) is the simplest multilead stem. Seals of this kind formed part of the vacuum envelope of nearly all receiving valves manufactured before 1940. They are now used in the manufacture of sub-miniature valves. The copper-clad leads are in one piece, as are those in sub-miniature versions of the envelope of (a).

The bulb is sealed to the base by flame heating with the electrode assembly in position. Seals must therefore be placed so that there is the minimum risk of oxidation of the assembly or of volatile material being deposited on it. Where extremely long life and high reliability are required, nitrogen or a hydrogen-nitrogen mixture may be blown through the envelpe while the seal is made. The exhaust tube is also flame sealed, and care must be taken that an adequately strong *pip* is formed when the seal is complete. The temperature and shape of the sealing flames must therefore be kept under close control during manufacture.

Apart from their vulnerability to impact, glass envelopes have the disadvantage that they can be used only where the heat dissipated therein can be lost by radiation without an unacceptable rise in temperature of the glass. Replacing the glass by silica increases the permissible heat dissipation many times, but the silica valve is limited in its applications by the need to use graded seals to the electrode supports. These can be made only by hand, and are therefore expensive in manufacture. Moreover, they can be used only in comparatively simple structures where support rods sealed through the envelope locate the electrodes, which must therefore be comparatively widely spaced, and so limit the valves to use at frequencies below a few tens of Mc/s. High power valves are therefore built with the anode forming part of the vacuum envelope and so accessible for air blast or water cooling.

Metal-glass Envelopes

Apart from the ease with which they can be cooled, tubes with metal-glass envelopes have the advantage that low-inductance connections can be made to the electrodes by discs or cylinders forming part of the vacuum enclosure. The greater part of the envelope of a microwave tube is metal, glass being used only in the cathode stem and the r.f. output window. The effects of thermal expansion are accommodated by arranging the seals so that the metal parts impose only a compression strain on the glass, or by using a ductile metal which "flows" and so relieves the strain on the glass, or one with an expansion coefficient matched to that of the glass up to the annealing temperature.

Copper-glass Envelopes

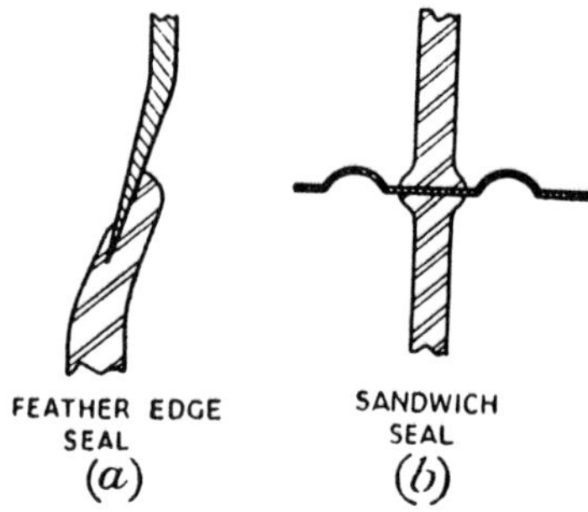

Copper to glass seals.
(a) Featheredge or Housekeeper seal.
(b) Sandwich seal.

Fig. 4.3

In these units, the strain on the glass is taken up by deformation of the copper, which has a low yield point and high ductility. There are two basic forms, the featheredge or Housekeeper seal (Fig. 4.3 (*a*)), and the sandwich seal (*b*), in which a thin lamina of copper is sealed between two glass cylinders. In both cases the seal is made between the glass and a thin film of red copper oxide on the surface of the metal.

The featheredge seal is used for the anodes of large transmitting valves. For adequate relief of strain the edge must be rolled down to a thickness less than 0·002″ and the taper to not more than 3°. The seal is made to the inside of the flare where the copper thickness is less than 0·05″, the outer surface of the larger seals being glassed to minimise oxidation of the metal by the heating flame. The sandwich seal is used in disc-seal triodes and other low-power valves for use at very high frequencies. The discs, preferably less than 0·03″ thick, are crimped to reduce their radial stiffness and sealed between cylinders of glass of equal diameter by r.f. induction heating. Two glass parts must be used for each seal so that flow of the copper occurs at both surfaces. Flow at one only causes tensile stress normal to the interface, and so failure of the seal. Either process is facilitated by a layer of copper borate on the metal surfaces. This is formed by heating the copper parts to red heat and immersing them in a hot concentrated solution of sodium borate. Seal of this kind are limited in their application by mechanical weakness, which results from the need to use thin ductile metal. They are therefore being replaced by seals using thicker sections of an alloy having its expansion more closely matched to that of the glass.

Alloy-glass Envelopes

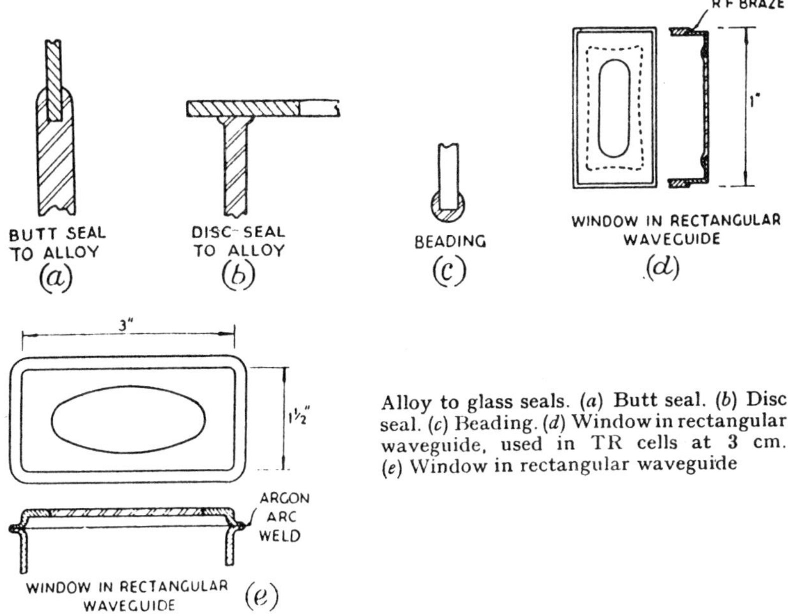

Alloy to glass seals. (*a*) Butt seal. (*b*) Disc seal. (*c*) Beading. (*d*) Window in rectangular waveguide, used in TR cells at 3 cm. (*e*) Window in rectangular waveguide

Fig. 4.4

The alloys used in the construction of small transmitting valves and microwave valves are those of iron, nickel, chromium and cobalt having different compositions according to which thermal expansion characteristic is required. Two of the most commonly used have expansions matched to those of soda lead and potash borosilicate glasses respectively.

No featheredge is required, and seals of the types shown in Fig. 4.4 can be made without difficulty provided the coefficients of thermal expansion of the alloy and glass are properly controlled.

The seal is usually made to a thin film of oxide which is formed by heating the metal parts to above 650°C in air before sealing. For some purposes, particularly where low r.f. loss is required, metal parts may be plated with silver or copper before being used. The deposited layer must be free from contaminants and strongly adherent to the base metal. This is usually much more difficult to achieve than the glass to metal bond, even though the plated parts are fired in a reducing atmosphere before the seal is made. Seals are made either by flame or eddy current heating. Thorough wetting of the alloy by the glass is essential to good sealing. This implies that the glass must flow over the surface as the seals are made. Allowance has therefore to be made for shortening of glass parts during the process, and the sealing fixtures designed so that there is freedom for motion of the metal parts normal to the axis of the assembly without loss of accuracy in the completed structure.

Windows for sealing waveguides must be made from a glass having low dielectric loss. These normally have lower thermal expansion than the sealing alloys in the range of temperature used in processing, and so a compression seal is used. The shape of the window must be such that compression in one direction does not produce tension in another. Circular windows fulfil this requirement best and are preferred to all others for handling high power. The window of Fig 4.4(*d*) is used in duplexer components with mild-steel bodies which are copper or gold plated to obtain the required r.f. characteristics. In the larger sizes, or where the window may be subject to thermal strain due to heating of the glass through dielectric loss, the elliptical window of (*e*) is preferred.

The main disadvantage of alloy-glass envelopes is that unless local cooling is used to protect them, the glass parts limit the temperature at which the assembly may be outgassed to below 450°C. This results in lower initial activity of the cathode than otherwise might be obtained.

5. PRODUCTION OF HIGH VACUA

Introduction

An electrode system within an envelope does not become an electron device until the proper environment for the moving electron has been created, and the cathode made capable of emitting electrons in great quantity at its chosen operating temperature. It will operate correctly only for the time that the environment remains unchanged or the cathode retains its activity. The high cost of assembly and processing of a high power microwave valve cannot be borne economically by the user unless it has long life. Consistency of characteristics, long life and high reliability are essential to the economic operation of communications and other equipment containing large numbers of general-purpose valves, even though their individual cost is low. High standards of vacuum technique are therefore essential to the successful manufacture of any electron device.

The exhaust processes for completing an electron device differ according to whether it is required in large or small numbers. In the first case, they must be done by machinery designed for mass production, in which the tube is carried in steps to a number of stations, stopping long enough at each for one of a series of operations to be completed before it passes to the next. In such a system, no individual attention to a particular tube is possible. The quality of the product depends entirely on the ability of the mechanism to maintain its adjustment over a long period. This in turn is determined by its tolerance of variations in ambient conditions, in mains supply voltage and in mechanical dimensions of the valve being manufactured. In the second case, the device is usually fixed in position and processed by moving into position the oven and other apparatus. The quality of the product then depends to a greater extent on the skill of the operator. The basic principles of design and operation of the exhaust machinery are, however, the same in both cases. The object is to remove all unwanted material from the tube in the shortest possible time, and to ensure that there are no deleterious after-effects.

The Exhaust Process

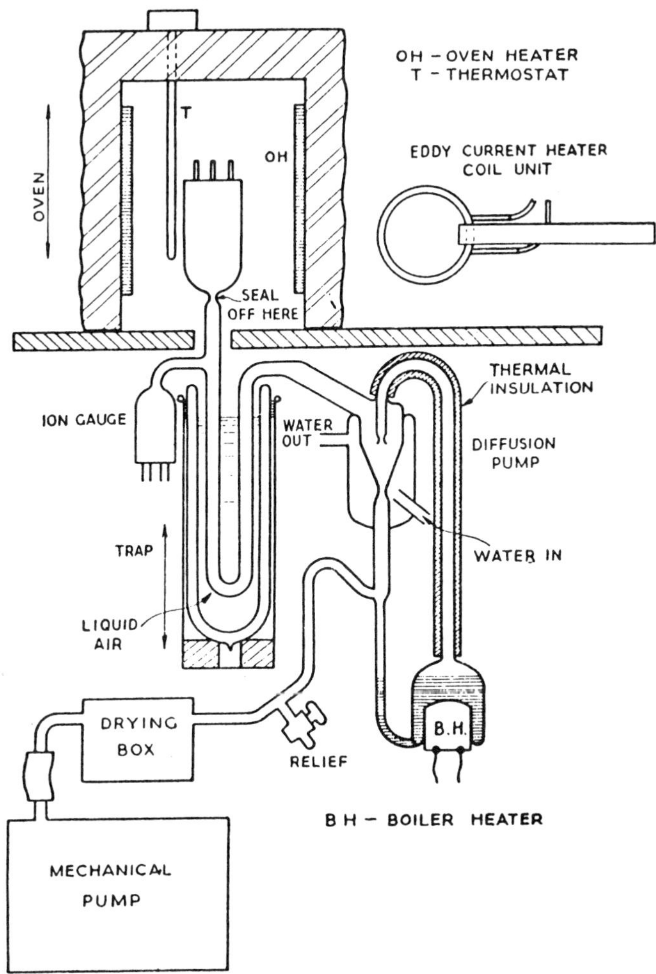

Bench-pumping system.

Fig. 5.1

The arrangement of a typical *bench pump* installation on which tubes are processed singly or in batches up to six at a time is shown in Fig. 5.1. It comprises:

(*a*) A motor-driven mechanical *fore pump* to *rough out* the system, i.e. to reduce the pressure to about 10^{-3} mm Hg.

(*b*) A *diffusion pump* to produce the ultimate vacuum, which is typically below 10^{-6} mm Hg.

(*c*) A *trap* to prevent vapour from the pump entering the valve being processed.

(*d*) An *ion* gauge to measure the pressure.

(*e*) A thermostatically controlled oven in which the valve is *baked out* for an adequately long time, and

(*f*) An eddy current heater coil unit, used for *outgassing* electrodes within a glass envelope or *firing* a getter.

The unit illustrated, which is typical of those used for pumping small transmitted valves, is built of hard glass. Larger systems may be built more conveniently of copper or steel. Individual components vary according to the choice of exhaust procedure. For example, oil-diffusion pumps with charcoal or water-cooled baffle traps are used in place of mercury vapour pumps when it is necessary to leave the system running unattended for long periods. The design of the pump depends upon the ultimate vacuum and the speed required. Where high baking temperatures are used for metal-envelope valves, the oven may be replaced by gas burners heating the body direct.

The vacuum processes used in bench-pumping an electron device are typically as follows. They are not necessarily performed on the same equipment.

(1) Metal piece-parts are furnaced in vacuo. Parts for microwave tubes are often given additional furnacing in hydrogen so that the pores are first cleansed of contaminants and then filled with a gas which is easily removed by raising the temperature. This minimises the risk of their later contamination by material more difficult to eliminate.

(2) In special cases, e.g. high-power microwave tubes, the cathode may be activated by means of an auxiliary anode and left sealed off in a container pending assembly into the remainder of the structure. Other sub-assemblies may be vacuum furnaced and stored in sealed-off containers.

(3) The completed valve is connected by its exhaust tube to the pumping system and the mechanical pump switched on to bring the pressure down to about 10^{-3} mm Hg.

(4) The oven is placed in position and the valve baked on the pump until the pressure is again reduced to 10^{-3} mm.

(5) The diffusion pump is switched on to bring the pressure down to near the ultimate value.

(6) In glass-envelope valves, the cathode temperature is raised but kept below that required to decompose the carbonate coating if such is used, and the getter and electrodes outgassed by means of the ECH coil.

(7) When the pressure has returned to below 10^{-4} mm Hg, the cathode temperature is raised to the *breakdown* value with the other electrodes held above their outgassing temperature by the ECH.

(8) When the pressure has been reduced to the required value (below 10^{-6} mm Hg with the cathode cold), the final seal is made by heating at

the constriction and drawing the valve away from the pumping tube.

(9) The getter, if any, is heated by ECH to volatilise the active material on to an area of the envelope well clear of the electrode system.

(10) The cathode may be activated by drawing current from a power supply with poor regulation so that the permissible anode dissipation is not exceeded as the cathode emission rises.

The time taken for exhaust depends on the complexity of the structure and the materials used in its fabrication. A radiation-cooled small transmitter tube built largely of molybdenum and tantalum may take up to an hour. Because of its copper body, a magnetron may take a few hours in all. Receiving valves pumped on rotating machinery, with 6 seconds spent at each of 24 stations, are processed in a slightly different way because of the short time available for processing and the low pumping speed which results from the use of a constricted exhaust tube and long piping back to the pump, which in some cases is a two-stage mechanical scavenger giving an ultimate pressure around 10^{-4} mm Hg without a diffusion stage. Outgassing of the electrodes begins after baking and before breakdown of the cathode and continues during the major part of the cycle, which is completed by firing the getter to produce the ultimate vacuum before sealing off. The cathode must not be broken down at pressures above 10^{-3} mm Hg.

Getters

Getters are fitted in electron devices when it is not possible to ensure that the pumping system gives the required ultimate vacuum, or that operation of the valve will not result in slow evolution of gas. The mass-produced general-purpose valve used in radio and communications equipment is an example of the first, the radiation-cooled small transmitting valve of the second. Getters are either films deposited on the inside of the vacuum envelope, or suitable metal surfaces operated at the correct temperature.

Barium is the material most commonly used as a deposited film getter. It is conveniently handled as a core within an iron wire forming part of a loop which is heated by ECH, the wire being ground away on one side for easy release of the volatilised material (see Fig. 5.2 (*a*)).

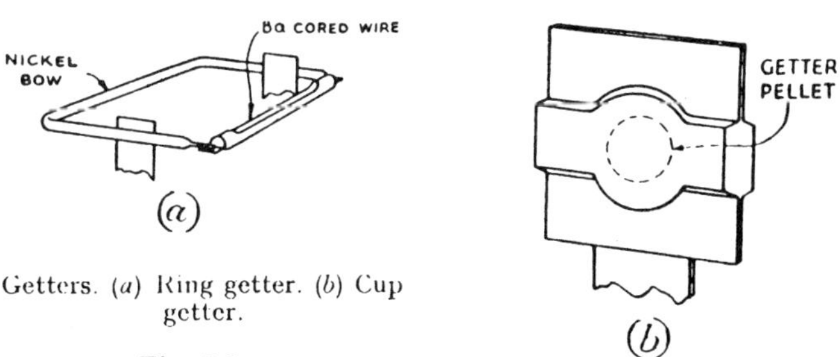

Getters. (*a*) Ring getter. (*b*) Cup getter.

Fig. 5.2

It is sometimes used as a pellet retained by nickel gauze on an electrode which will be raised to a suitable temperature during exhaust, or contained in a cup of nickel shaped so that the deposit is confined to a selected region of the envelope (see Fig. 5.2 (*b*)). The getter is *fired* as the last operation on the pump or immediately after seal off, and combines with the residual gas within the envelope to produce a hard vacuum. The film remains active and continues to absorb gas as it is evolved during operation of the valve, so maintaining the condition necessary for satisfactory operation.

Glossary

Long Line Effect

When a magnetron is connected to a load with a reflection coefficient greater than 0.1 through a waveguide many wavelengths long (for example, in airfield surveillance radar systems where the aerial is at the top of a tripod mast and reflection of the transmitted signal may occur at rotating joints or from the radiating element itself), the reactive component of the impedance seen by the magnetron can vary significantly with frequency. Under these conditions, there may be more than one permissible frequency of operation because of the cumulative effect of the reactance changing the frequency of the magnetron, so causing a change in the reactance, and so on until the reactance is reduced to zero. The magnetron can then jump at random from one frequency to the other and back. This 'long line effect' causes a degradation in performance of the radar system because the receiver cannot follow the random change in frequency of the transmitted signal. The long line effect can be reduced by decreasing the coupling between the resonator system of the magnetron and the waveguide. This results in some loss of power output, and so a compromise has to be made between available power output and 'pulling figure', which determines the maximum permissible length of the transmission system. [In a typical X-band magnetron, the coupling is arranged to give a pulling figure around 15 MHz, which limits the length of the waveguide to 5 metres when the reflection coefficient of the load is 0.2]. Alternatively, power from the magnetron can be fed into the waveguide through a directional coupler, which eliminates the effect of reflections of the transmitted signal.

Moding, Mode Splitting

'Moding' describes the generation by a magnetron of outputs at more than one frequency. This may occur during the rise of the DC input pulse, or be manifested as 'mode splitting', a random and repetitive shift from one frequency to another nearby and back when the input voltage has risen to the operating level.

Pulling Figure

'Pulling Figure' is a measure of the frequency stability of a magnetron when it is connected to a load whose impedance contains a significant reactive component which can vary in phase and amplitude. It is measured by connecting a resistive load with a specified reflection coefficient (usually 0.2) to the magnetron through a short length of waveguide and measuring the difference between the maximum and minimum output frequencies when the electrical length of the waveguide is altered by at least one half-wavelength. (This is accomplished by varying the intrusion of a blade of very low loss dielectric into the guide through a longitudinal slot in its broad face.)

'Q'

The 'Q' of a tuned circuit is defined as the ratio at resonance of either the inductive (ωL) or the capacitative ($1/\omega C$) reactance to the total series resistance r. (In practice the resistance of the capacitor can be ignored in comparison with that of the inductor). It can be shown that the fractional difference in frequency (f1-f2)/f0 between the points on the resonance curve at which the response falls to the 'half power' level (70.7 per cent of the peak voltage amplitude) is 1/Q, and that the dynamic resistance of a shunt tuned circuit at resonance (L/Cr) can be written as ωLQ.

Thus a high Q circuit has a very high narrow peak. The response at frequencies on either side of the peak is increasingly independent of Q.

Radar Equation

The radar equation gives an approximate value of the power P_r received after reflection from a target at range R when it is illuminated by a radar system radiating a peak power P_t through a directional aerial system in which the transmitting and receiving antennae have the same polar diagram and area A. [In common T and R systems there is only one antenna]. Neglecting losses in the protective housing of the antennae or during transmission and reflection

$$P_r = \frac{P_t G}{4\pi R^2} \cdot \frac{S}{4\pi R^2} \cdot KA$$

where S is the effective area of the target, G is the gain of the transmitting aerial and K is a factor with a value between 0.5 and 0.9 depending upon the configuration of the antennae. $P_t G/4\pi R^2$ is the power per unit area impinging on the target; $S/4\pi R^2$ is the factor by which the illuminating power is reduced by the time the reflected signal has returned to the receiving system. 'Gain' is the ratio of the power in the cross-section of the beam to that which would inpinge upon an equivalent area of a uniformly illuminated sphere surrounding the transmitting antenna. It can be shown that

$$G = \frac{4\pi KA}{\lambda^2}$$

where λ is the wavelength of transmission. Thus the radar equation can be written as

$$P_r = \frac{K^2 A^2}{4\pi\lambda^2} \cdot \frac{P_t S}{R^4}$$

At 10 cm wavelength, using a paraboloid dish antenna with a typical area of 3 m²,

$$P_r \approx \frac{20 P_t S}{R^4} \quad \text{or} \quad R \approx \sqrt[4]{20S.\frac{P_t}{P_{rm}}}$$

where P_{rm} is the minimum signal which the receiver can detect. S is of the order fractions of a square metre for a marker buoy or a periscope, between 10 and 20 m^2 for a small aircraft or a submarine awash, and about ten times that amount for a large aircraft or a small ship.

Wavebands (approximate)
    K   1.2–1.6 cm (around 20 GHz)
    X   3.0–3.3 cm (9 to 10 GHz)
    S   9.0–11 cm (2.7 to 3.3 GHz)
    Metric   (For the purposes of this text)   40 MHz to 1 GHz

Waveguide

A rectangular metal (usually copper) tube, through which microwave energy can be transmitted with very low attenuation, provided the half-wavelength of the transmitted signal is a large fraction (but typically less than 0.9) of the longer transverse dimension of the tube. In simple terms, the waveguide can then be considered to be a twin transmission line with its input and output terminals at the centres of the broad sides, shunted on each side by stubs slightly longer than one quarter-wavelength which prevent loss of energy by radiation. A series connection is made by attaching the side arm of a T to the broad face of the guide. A parallel connection is made by attaching it to the narrow side. Energy can also be coupled inductively into or out of the guide through apertures in the narrow side (whose dimension is approximately 0.4 of that of the broad face).

Numerical index

Subject index